Kanban for
the Shopfloor

Kanban for the Shopfloor

CREATED BY

The Productivity Press
Development Team

Productivity Press
NEW YORK • NEW YORK

Additional copies of this book are available from the publisher. Discounts are available for multiple copies through the Sales Department (888-319-5852). Address all other inquiries to:

Productivity Press
444 Park Avenue South, Suite 604
New York, NY 10016
United States of America
Telephone: 212-686-5900
Fax: 212-686-5411
E-mail: info@productivitypress.com

Cover art direction by Stephen Scates
Cover illustration by Gary Ragaglia
Content development by Diane Asay, LeanWisdom
Page design and composition by William H. Brunson, Typography Services
Printed and bound by Malloy Lithographing, Inc. in the United States of America

Library of Congress Cataloging-in-Publication Data

Kanban for the shopfloor / created by the Productivity Press Development Team.
      p.   cm. — (Shopfloor series)
   Includes bibliographical references.
     ISBN 1-56327-269-5
     1. Production management.   2. Costs, Industrial.   I. Productivity Press.
   Development Team.   II. Series.
   TS155 .K287   2002
   658.5—dc21

                                                                2001007647

06 05   5 4

Contents

Chapter 2. The Basics of Kanban: Functions, Rules, and Types of Kanban   **11**

Publisher's Message

Kanban for the Shopfloor was written to give you a powerful tool to help you "lean" your shopfloor. The primary benefit of kanban is to reduce overproduction—the most critical of the seven deadly wastes. Kanban produces exactly what is ordered, when it is ordered, and in the quantities ordered. It has been called the nervous system of lean production because it manages production just as our brains and nerves manage our bodies. As you apply kanban in your shop, it will become the communication system that keeps everything integrated and in harmony.

This book will give you an orientation to the complex topic of kanban in an organized and easily-to-assimilate manner. The "Getting Started" section will suggest learning strategies and give you an overview of each chapter. Chapters I through 5 will give you practical information as well as helpful how-to steps. Throughout the book, you will be asked to reflect on questions to help you apply kanban to your unique shopfloor circumstances. Numerous illustrations will reinforce the text. A "Summary of the Steps for Kanban Implementation" and resources for continued learning will be provided for you in Chapter 6. This book will tell you what you need to know to begin your kanban efforts, as well as give you valuable information for maintaining and improving upon your initial efforts.

Kanban works best if your company has committed to a system of pull production. Lean methods such as one-piece flow, cellular manufacturing, quick changeover, autonomous maintenance, and 5S will increase the benefits you will receive from your kanban implementation. We have described these lean methods in our Shingo-Prize-winning Shopfloor Series, which is all about improving performance on the factory floor. This book is the first addition to that prize-winning series. Watch for further titles, coming out soon, on kaizen, standard work, and pull production.

Acknowledgments

The development of *Kanban for the Shopfloor* has been a team effort and we wish to thank the following people. Judith Allen, Vice President of Product Development, spearheaded this project. Special thanks to Diane Asay of LeanWisdom for shaping and writing the content, using material from the Productivity archives. Art Director Stephen Scates created the cover design and concept, with cover illustration provided by Gary Ragaglia of Metro Design. Mary Junewick was the project manager and copyeditor. Lorraine Millard created the numerous illustrations. Guy Boster created the cartoons. Typesetting and layout was done by Bill Brunson of Typography Services. Toni Chiapelli was our proofreader. Michael Ryder managed the print process. Finally, thanks to Lydia Junewick and Bettina Katz of the marketing department for their promotional efforts.

We are very pleased to bring you this addition to our Shopfloor Series and wish you continued and increasing success on your journey to lean.

Sean Jones
Publisher

Getting Started

The Purpose of This Book

Key Point

Kanban for the Shopfloor was written *to give you the information you need to participate in implementing this important lean manufacturing approach in your workplace.* You are a valued member of your company's team; your knowledge, support, and participation are essential to the success of any major effort in your organization.

You may be reading this book because your team leader or manager asked you to do so. Or you may be reading it because you think it will provide information that will help you in your work. By the time you finish Chapter 1, you will have a better idea of how the information in this book can help you and your company eliminate waste and serve your customers more effectively.

The Basis of This Book

BACKGROUND INFO

This book is about an approach to production scheduling and communication between processes to avoid overproduction and overstocking. The methods and goals discussed here are an important part of the lean manufacturing system developed at Toyota Motor Company. Since 1979, Productivity, Inc. has brought information about these approaches to the United States through publications, events, training, and consulting. Today, top companies around the world are applying lean manufacturing principles to sustain their competitive edge.

Kanban for the Shopfloor draws on a wide variety of Productivity's resources. Its aim is to present the main concepts and techniques of the kanban system in a simple, illustrated format that is easy to read and understand.

Two Ways to Use This Book

There are at least two ways to use this book:

1. As the reading material for a learning group or study group process within your company.

2. For learning on your own.

The management of your company may decide to use this book to design its own learning group process. Or, you may read this book for individual learning without formal group discussion. Either way, you will learn valuable concepts and techniques to apply to your daily work.

How to Get the Most Out of Your Reading

Becoming Familiar with This Book as a Whole

There are a few steps you can follow to make it easier to absorb the information in this book. Take as much time as you need to become familiar with the material. First, get a "big picture" view of the book by doing the following:

How-to Steps

1. Scan the "Table of Contents" to see how *Kanban for the Shopfloor* is arranged.

2. Read the rest of this introductory section for an overview of the book's contents.

3. Flip through the book to get a feel for its style, flow, and design. Notice how the chapters are structured and glance at the pictures.

Becoming Familiar with Each Chapter

After you have a sense of the structure of *Kanban for the Shopfloor*, prepare yourself to study one chapter at a time. For each chapter, we suggest you follow these steps to get the most out of your reading:

How-to Steps

1. Read the "Chapter Overview" to see what the chapter will cover.

2. Flip through the chapter, looking at the way it is laid out. Notice the bold headings and the key points flagged in the margins.

3. Now read the chapter. How long this takes depends on what you already know about the content and what you are trying to get out of your reading. Enhance your reading by doing the following:

- Use the margin assists to help you follow the flow of information.

- If the book is your own, use a highlighter to mark key information and answers to your questions about the material. If the book is not your own, take notes on a separate piece of paper.

- Answer the "Take Five" questions in the text. These will help you better absorb the information by reflecting on how you might apply it to your own workplace.

4. Read the "Summary" at the end of the chapter to recap what you have learned. If you read something in the summary you don't remember, find that section in the chapter and review it.

5. Finally, read the "Reflections" questions at the end of the chapter. Think about these questions and write down your answers.

How a Reading Strategy Works

When reading a book, many people think they should start with the first word and read straight through until the end. This is not usually the best way to learn from a book. The steps just described for how to read this book were a strategy for making your reading easier, more fun, and more effective.

Key Point

Reading strategy is based on two simple points about the way people learn. The first point is this: *It's difficult for your brain to absorb new information if it does not have a structure to place it in.* As an analogy, imagine trying to build a house without first putting up a framework.

Like building a frame for a house, you can give your brain a framework for the new information in the book by getting an overview of the entire contents and then flipping through the materials. Within each chapter, you can repeat this process on a smaller scale by reading the overview, key points, and headings before reading the text.

Key Point

The second point about learning is this: *It is a lot easier to learn if you take in the information one layer at a time, instead of trying to absorb it all at once.* It's like finishing the walls of a house: First you lay down a coat of primer. When it's dry, you apply a coat of paint, and later a final finish coat.

Using the Margin Assists

As you've noticed by now, this book uses small images called *margin assists* to help you follow the information in each chapter. There are six types of margin assists:

Background Information Sets the stage for what comes next

Key Term Defines important words

Key Point Highlights important ideas to remember

Example Helps you understand the key points

New Tool Helps you record data or apply what you have learned

How-to Steps Indicates the sequence for improvement action

An Overview of the Contents

Getting Started (pages xi–xvi)

This is the section you have been reading. It has already explained the purpose of *Kanban for the Shopfloor* and how it was written. It also shared tips for getting the most out of your reading. Now, it will present a brief description of each chapter.

Chapter 1. Introducing Key Terms and Benefits of Kanban (pages 1–10)

Chapter 1 introduces and defines the kanban system and the kinds of waste it helps eliminate, and defines the pull system as a framework for the rest of the book. It discusses what is needed to be successful in implementing kanban and explains how the kanban approach benefits companies and their employees.

Chapter 2. The Basics of Kanban: Functions, Rules, and Types of Kanban (pages 11–27)

Chapter 2 describes the primary functions of the kanban system and gives the rules for using kanbans—to withdraw from the upstream process, to produce what has been withdrawn, etc. The types of kanban are also explained.

Chapter 3. Phase One: Scheduling Kanban (pages 29–49)

Chapter 3 highlights the key steps for creating a production schedule in the kanban system. It describes how to determine the number of kanbans needed, and how to calculate takt time and balance the lines so that everyone is working in harmony with customer orders. Load leveling or *heijunka* is defined and described as well.

Chapter 4. Phase Two: Circulating Kanban (pages 51–65)

Chapter 4 covers essential methods that support shopfloor implementation of the kanban system. It discusses when to produce, when to pull, and how to create kanbans. Steps are given for circulating kanbans and rules are given to support transport and production. Supermarkets, water beetles, and milk runs are also described.

Chapter 5. Phase Three: Improving with Kanban (pages 67–87)

Chapter 5 discusses the Kanban system as an improvement method. By reducing the number of kanbans, problems are revealed and work-in-process stocks are reduced to a minimum. Kanban as a visual control system is explored and the use of *triangle* kanbans and other kanbans for special circumstances is described. The chapter concludes with how your kanban system relates to suppliers.

Chapter 6. Reflections and Conclusions (pages 89–95)

Chapter 6 presents reflections on and conclusions to this book. It includes an implementation outline of the kanban system. It also describes opportunities for further learning about the kanban system and related techniques.

Chapter 1

Introducing Key Terms and Benefits of Kanban

Supply info	Part info	Customer info
	Part # **52107**	User processes **Small parts DSG**
Raw material code **4" x 4' cants**	Description **3/8" board x 4'**	Storage locations **C-12**
Raw material location **Shed 1 - B6**	Quantity **400/skid**	Kanban #/Issue date **#4 - 3/18/95**

Figure 1-1. A Typical Kanban Card

What Is Kanban?

The kanban system determines the production quantities in every process. It has been called *the nervous system* of lean production because it manages production just as our brains and nerves manage our bodies. The primary benefit of the kanban system is to reduce overproduction; and its aim is to produce only what is ordered, when it is ordered, and in the quantities ordered.

Key Term

Key Point

In Japanese, the word "kanban" means "card" or "sign" and is the name given to *the inventory control card used in a pull system*. It is essentially a work order that also moves with the material. Each card or kanban identifies the part or subassembly unit and indicates where each came from and where each is going. Used this way, kanban acts as *a system of information that integrates the plant, connects all processes one to another, and connects the entire value stream to the customer demand harmoniously*.

The Pull System and Waste Reduction

In the kanban system the upstream process produces only enough units to replace those that have been withdrawn by the downstream process. Workers in one process go to the preceding process to withdraw the parts they need. They do this only in the quantities and at the time when the units are needed. The start of this withdrawal system begins with a customer order. This is called a *pull system*.

Key Term

2

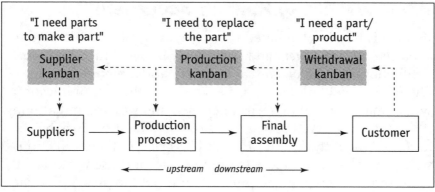

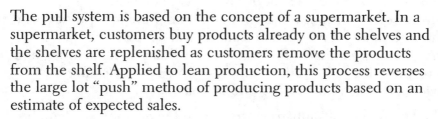

Figure 1-2. Pull Production with Kanban

The pull system is based on the concept of a supermarket. In a supermarket, customers buy products already on the shelves and the shelves are replenished as customers remove the products from the shelf. Applied to lean production, this process reverses the large lot "push" method of producing products based on an estimate of expected sales.

A pull system creates *flexibility* on the production floor so that exactly what has been ordered will be produced, when it is ordered, and only in the quantities ordered. In this way, it is possible to *eliminate overproduction*—the most critical of the seven deadly wastes. The ultimate *conceptual goal* is *zero kanban*, which eliminates work-in-process inventory. In other words, your customer order becomes the signal for a pure continuous flow. It is the *ideal* that the spirit of continuous improvement in a pull system always strives toward. Reducing kanbans will be discussed in Chapter 5.

TAKE FIVE

Take five minutes to think about these questions and to write down your answers:

1. What are two other words for kanban and how is a kanban used?
2. What are some advantages of a pull system?

What Will Make Kanban Successful?

Kanban is best implemented when a company has committed to a *pull production system* and has already implemented small-lot production through *one-piece flow* and *cellular manufacturing*. Once these methods are in place, kanban is the communication system that keeps all the cells integrated and working in harmony. If kanban is deployed only in parts of the plant there may be confusion between the "push" and "pull" aspects of the production system. Kanban will expose the problems that cause waste, such as overproduction, and if a pull system is not the plant's ultimate goal it may become difficult to eliminate these problems. If a company has wide fluctuations in demand that cannot be smoothed (such as seasonal products) and therefore will not benefit from small-lot production, kanban will be less effective and may be inappropriate. To learn about cellular manufacturing and one-piece flow see Productivity's shopfloor book, *Cellular Manufacturing: One-Piece Flow for Workteams*.

As the number of kanbans in the system is gradually reduced, the first problems that surface are usually changeover bottlenecks. Improvement methods must be used immediately to reduce changeover times so that takt time can be reestablished and a mixed, small-lot production flow can continue to be regulated by kanbans. If methods for reducing changeover times are not being practiced it will be impossible for the plant to respond to customer demand, which is the key purpose of kanban and the pull system. Quick changeover is discussed fully in Productivity's shopfloor book, *Quick Changeover for Operators: The SMED System*.

Autonomous maintenance is another critical element to insure a successful pull production system. Keeping machines operating, planning scheduled maintenance, and all the other elements of total productive maintenance ultimately will be necessary for kanban to function optimally. Our *Autonomous Maintenance for Operators* shopfloor book will help you implement this lean method.

Kanban is an *advanced visual control system* and depends on the discipline and understanding that grow from implementing the 5S. The steps toward creating a *visual workplace*—beginning with 5S to put the workplace in order, establish visual displays, and support continuous improvements (which are initiated by each and every operator)—will be an important foundation for your pull system implementation. The steps of 5S are fully discussed in Productivity's *5S for Operatars: 5 Pillars of the Visual Workplace*. Chapter 5 of this book includes a discussion of kanban as a visual control system.

TAKE FIVE

Take five minutes to think about these questions and to write down your answers:

1. How many of these lean production methods do you currently use in your plant?
2. What else can you do to prepare your workplace for the kanban system?

Integrating Kanban with MRPII

Many books have been written discussing ways to integrate a material requirements planning system with kanban. We will not address this approach here. MRPII is a computerized system for estimating future production needs rather than responding to customer demand as it occurs. In other words it was designed for use in "push" systems. Where some companies may choose to move gradually toward a pull system by integrating their existing MRPII method with kanban, the focus of this book is to show the kanban system in its purest form as the mechanism for implementing true "pull" production.

Pilot or Plantwide Implementation

It is important to decide where kanban will be implemented—throughout the plant or only in pilot areas. Remember, kanban is a system that integrates the processes in a plant and ties them to the customer. Choosing only a pilot area may limit the ultimate effect as well as the purpose of kanban itself.

However, it is possible to begin with kanban in a pilot area, even if continuous flow manufacturing has not been implemented. In this case, however, kanban will serve to reveal the problems in the flow of production. As the number of kanbans is reduced, long changeover requirements, transportation delays, machine downtime, and cumbersome WIP piles will interrupt production. Kanban will not solve these problems; it will only make them visible.

It will then be necessary to implement the other methods of lean production—5S, SMED, autonomous maintenance, and plant layout to support cells and one-piece flow—before kanban can become what it was meant to be: *the communication mechanism that supports pull production.*

On the other hand, if you have implemented 5S, quick change-over, and autonomous maintenance, and if you are committed to establishing a pull system, then we highly recommend you establish kanban plantwide. At this point in your improvement process kanban will integrate and synchronize all the production processes and create a rhythm on the factory floor—the rhythm of *takt time,* the pulse of customer demand. Kanban will reveal the deeply embedded bottlenecks that otherwise might have remained hidden, and lean production will become a reality.

How Will Kanban Change What You Are Doing Now?

We are all taught to be efficient—the more we are able to produce the better workers we are. This is what we have always been told. We have trained ourselves to work hard, to live up to that old

Figure 1-3. Overproduction—Producing Just Because You Can

law of the workplace that "more is better." In a lean factory with a pull production system using kanban, this idea *must go.*

With kanban "more is just more," and this need to produce just because you can leads to the worst of the seven deadly wastes—overproduction. *In kanban, workers produce only when signaled.* Kanban is the signaling system and what is signaled comes from the downstream process, starting with a customer order.

In this book you will learn how to create the signals that you need for your process—production kanbans and withdrawal kanbans. You will discover the value of leveling production and balancing the lines or cells so that *the rhythm of production becomes synchronized with the demand of the customer.*

TAKE FIVE

Take five minutes to think about these questions and to write down your answers:

1. How do you know when and what to produce now?
2. How would a kanban system change this?

What Are the Benefits of Kanban?

We have already mentioned some of the benefits of kanban and pull production. Let's look specifically at how they will benefit you and your company.

Kanban and Your Company

Kanban will help your company:

1. Eliminate overproduction: the #1 waste

2. Increase flexibility to respond to customer demand

3. Coordinate production of small lots and wide product variety

4. Have a simplified procurement process

5. Integrate all processes and tie them to the customer

Kanban and You

Kanban will help you:

1. Connect information with the part or product

2. Find simple, visual, replenishment information

3. Find simple, visual, production instructions

4. Eliminate unnecessary WIP inventory

5. Uncover hidden waste in your process

TAKE FIVE

Take five minutes to think about these questions and to write down your answers:

1. Based on what you have read, can you see how a kanban system might benefit your company? If so, how?
2. Can you see how it might benefit you? If so, how?

In Conclusion

SUMMARY

Kanban has been called *the nervous system* of lean production because it manages production just as our brains and nerves manage our bodies. The primary benefit of the kanban system is to reduce overproduction; and its aim is to produce only what is ordered, when it is ordered, and in the quantities ordered. "Kanban" means "card" or "sign" and is the name given to *the inventory control card used in a pull system*. It is a work order that moves with the material. Each kanban identifies a part or sub-assembly unit and indicates where each came from and where each is going. Used this way, kanban is as *a system of information that integrates the plant, connects all processes one to another, and connects the entire value stream to the customer demand harmoniously.*

The upstream process produces only enough units to replace those that have been withdrawn by the downstream process. Workers in one process go to the preceding process to withdraw the parts they need—only in the quantities needed and when needed. The start of this withdrawal system begins with a customer order. This is called a *pull system.* It reverses the large lot "push" method of producing products based on an estimate of expected sales and makes it possible to *eliminate overproduction*—the most critical of the seven deadly wastes.

Kanban is best implemented when a company has committed to a *pull production* system and has already implemented small-lot production through *one-piece flow* and *cellular manufacturing.* *Quick changeover* (or SMED) and *autonomous maintenance* will also be essential for kanban to succeed. Kanban is an *advanced visual control system* and depends on the discipline and understanding that grow from implementing the 5S.

We recommend that you establish kanban plantwide to integrate and synchronize all the production processes and create a rhythm on the factory floor—the rhythm of *takt time,* the pulse of customer demand. Kanban will reveal the deeply embedded bottlenecks that otherwise might have remained hidden, and lean production will become a reality.

In kanban, workers produce only when signaled. Kanban is the signaling system and what is signaled comes from the downstream process, starting with a customer order. Kanban will help your company eliminate overproduction, increase its flexibility to respond to customer demand, coordinate production of small lots and wide product variety, have a simplified procurement process, and integrate all processes by tying them to the customer. Kanban will help you connect information with the part or product, find simple, visual, replenishment information and production instructions, eliminate unnecessary WIP inventory, and uncover hidden waste in your process.

REFLECTIONS

Now that you have completed this chapter, take five minutes to think about these questions and to write down your answers:

- What did you learn from reading this chapter that stands out as particularly useful or interesting?

- Do you have any questions about the topics presented in this chapter? If so, what are they?

- What additional information do you need to fully understand the ideas presented in this chapter?

Chapter 2

The Basics of Kanban: Functions, Rules, and Types of Kanban

The Differences Between Kanban and Conventional Ordering Systems

Why does inventory accumulate?

Reordering Point Method and Kanban

Key Point

The kanban system is based on an inventory management system called reordering point method. This is a statistical method that allows factories to reorder the same amount of parts or products each time. When inventory drops to a certain level—the reorder point—a new order is made to replace the used inventory. The reordering point method can be automated easily and it keeps inventory management costs down by reducing clerical work. However, it does not pay attention to changes in demand. In fact, this method is unsuitable where sharp fluctuations in demand exist.

Key Point

This is also true of kanban. Kanban requires a somewhat stable market environment. It is unsuitable for products that have large and unpredictable ups and downs. Even kanban cannot prevent shortages or gluts in such cases. Instead, *the kanban system minimizes waste by using level production,* which averages out the product models and volumes to be produced, and eliminates the need to produce in large lots.

Level production is discussed in detail in the next chapter. The pull system book in this Shopfloor Series describes in detail how to gradually lower the reorder point and cut supply lot sizes to a minimum. Level production depends on setup time reduction described in the *Quick Changeover for Operators* book, and on reductions in minimum inventory needed to deal with production instability. Kanbans become the means of visual control to keep this minimum inventory system going.

Though the kanban system grew out of the reordering point system and shares some of its characteristics, kanban is a great improvement over the earlier system in many ways. See Figure 2-1 to see the similarities and differences of the two systems. These benefits of the kanban system will be discussed in detail in the next two chapters.

		Reordering Point Method	**Kanban System**
Similarities		1. Enables inventory to be managed without paying attention to demand fluctuations 2. Not suitable when sharp demand fluctuations are typical 3. Helps keep inventory management costs down 4. Conducive to use in an automated reordering system	
Differences	**Information and goods**	Information and goods are kept separate from each other (inventory [=goods] is managed according to the warehouse entry/exit vouchers [=information])	Information (kanban) and goods are kept together
	Management	Requires constant inventory management (warehouse entry/exit management)	Does not require management
	Visual control	Does not enable visual control	Enables visual control
	Relationship with factory	Managed separately from the factory	Closely related to the factory and factory operations
	Relationship to improvement activities	None	Decreasing numbers of kanbans indicate the need for improvement

Figure 2-1. Similarities and Differences Between the Reordering Point Method and the Kanban System

Production Work Orders and Kanban

The "push" production system depends on *production work orders* to identify the type and quantity of production to be done at each process. Production work orders are generally used when upstream processes determine how and when goods are moved downstream and how they are controlled between processes. Production work orders are based on process-specific operation plans developed as part of the production schedule for the plant. Thus, even though production is still a series of processes, each process relates vertically to the production schedule, not horizontally to other processes.

Key Point

In the kanban system, *kanbans serve as the production order for the pull system.* They follow the goods and indicate what is to be withdrawn from the upstream process. As soon as a client orders a product, a work order is sent to the assembly line, which in turn orders parts from the process line. The process line orders the materials needed from procurement and so on. You can see that this is the reverse of production in the push system, which begins at parts procurement and moves downstream. *Order information in a pull system—the kanban—travels upstream from sales to*

13

assembly to procurement instead of downstream from planning and procurement to assembly to sales.

Since the downstream processes drive a pull system, originating with the customer order, there is great flexibility in relation to demand, and in-process waste also can be greatly reduced. In Figure 2-2 you can see the differences between a push and a pull system using kanbans.

MRP II

Material requirement planning is a unique and widely used computerized push system technique. It uses a concept called *time buckets*—a predetermined time period for the production of a set quantity of units based on lead-time data. In the pull system of

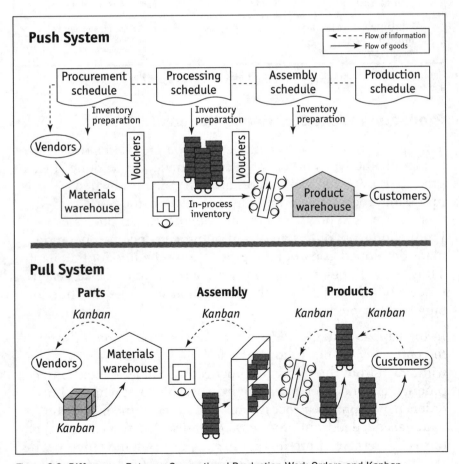

Figure 2-2. Differences Between Conventional Production Work Orders and Kanban

kanban this time bucket would be one day of production whereas in MRP it covers at least a week.

MRP depends on a master schedule that is rigorously maintained. Planned and actual production output are compared daily and discrepancies are remedied weekly through changes in the master schedule. Though the kanban system also has an overall production schedule it does not drive production but sets the stage for it. It is used to prepare the plant for the day's work, arranging materials and workers at each process. Comparison of plans and actual production is not necessary since production occurs in response to daily orders.

Key Point

Since kanbans flow backward from the final assembly upstream, only the final assembly line needs to know about changes in sequence or quantity. The entire plant's production will automatically respond to these changes as the kanbans reach each process accordingly.

However, kanban can be used in an MRP system as a dispatching tool within each time bucket. Also, for products that do not allow smoothing of production because of very short production runs or sharp fluctuations in demand, MRP may be more appropriate.

A Summary of Push versus Pull

In summary, the push system driven by production work orders imposes in-process inventory onto downstream processes. It tries to stubbornly fulfill the production schedule no matter what is happening downstream. The rigidity of the push system stems from the need to determine the schedule for the next week and to estimate future orders for the next several weeks, and it requires large lot sizes and long production cycle times. Process waste abounds (see Figure 2-3).

Once setup times and minimum inventory are reduced and order point is lowered so that small lot size and level production are possible, a pull system can be established that allows maximum flexibility in relation to customer orders and downstream production needs.

Figure 2-3. Overproduction Causing In-Process Waste

In a pull system the downstream processes determine the production demands of upstream processes. The downstream process "pulls" from the upstream process just those goods that are needed, when they are needed, and in the amounts needed. As products are pulled from one upstream process, it pulls from the previous process and, like a chain reaction, products and parts are pulled all the way back from raw material inventory and ultimately from outside suppliers.

Figure 2-4. Pull System Production—Production As Needed

TAKE FIVE

Take five minutes to think about these questions and to write down your answers:

1. What type of scheduling system is used at your plant?

2. How does a pull system integrate a plant?

Functions of Kanban

To Serve as the Autonomic Nervous System for Just-in-Time Production

Just as the autonomic nervous system signals the brain when certain conditions occur, so kanban is the communication system for lean production. It signals upstream processes when and what to produce and alerts them when problems or changes occur so that they can stop producing.

Key Point

Pickup and work order information: Kanbans serve as work orders; they are automatic directional devices. Kanbans provide two kinds of information:

1. What parts or products have been used and how many

2. Where and how parts or products are to be produced

Kanbans signal standard operations to be engaged at any time based on actual conditions existing in the workplace. They also prevent unnecessary paperwork in startup operations.

Elimination of overproduction waste: Since production only occurs when signaled by a downstream process, in-process inventory and transportation are kept at a minimum and overproduction does not occur.

To Improve and Strengthen the Factory

Kanbans remain attached to the goods that they define and consequently serve as visual controls.

Key Point

A tool for visual control: Since kanbans stay with the goods until the product is completed, they act as visual indicators of where production priorities exist and how operations are proceeding. Since they drive production, they are powerful visual controls of the process itself, determining when each process is to produce more and when it is to stop production.

Key Point

A tool for promoting improvement: Inventory hides problems. Too many kanbans indicate excess in-process inventory. By reducing the number of kanbans, problem areas will come out of hiding so that they can be improved. In this way, the kanban system becomes a valuable means to drive out waste and continually improve the production system.

TAKE FIVE

Take five minutes to think about these questions and to write down your answers:

1. What are the two primary functions of kanban? Why are they important?
2. What information do kanbans provide? Is this information that you need to know?

Rules of Kanban

The following rules must be observed if the kanban system's potential is to be fully realized:

> **Rule 1:** Downstream processes withdraw items from upstream processes.
>
> **Rule 2:** Upstream processes produce only what has been withdrawn.
>
> **Rule 3:** Only 100 percent defect-free products are sent to the next process.
>
> **Rule 4:** Level production must be established.
>
> **Rule 5:** Kanbans always accompany the parts themselves.
>
> **Rule 6:** The number of kanbans is decreased gradually over time.

Rule 1: Downstream Processes Withdraw Items from Upstream Processes

This rule transforms the idea of "supplying" to one of "withdrawing" and in one stroke solves the difficult problem of overstocking. The following steps must be followed for this rule to be effective:

- Make no withdrawals (nothing is transported) without a kanban.
- Withdraw only as many parts as the kanban indicates.
- A kanban must always accompany each item.
- Go from one process to the preceding one to withdraw parts.

This rule insures that only what has been sold will be made. This is not just the first rule in the kanban system but a crucial factor in lean production itself.

Rule 2: Upstream Processes Produce Only What Has Been Withdrawn

Only the exact quantity withdrawn by the subsequent process will be produced. This prevents overproduction by constraining the total flow of parts. It also holds in-process stock to a minimum so that parts must be made in the order they have been withdrawn to avoid shortages.

- Do not produce more than the number of kanban received.
- Produce in the sequence in which the kanban are received.

Rule 3: Only 100 Percent Defect-Free Products Are Sent to the Next Process

Quality is built in at each process. This is so important that some make it the first rule of kanban. Like Rule 1, this is also a defining feature of lean production itself.

Each process discovers and corrects its own defects. Machines must be able to stop when defects occur so that the problem can be remedied; and workers must be able to stop producing when problems arise. If defective products are not discovered until the next process has withdrawn them, do not modify the kanban. Instead, ask that the exact number of defective items be replaced in the next pickup.

Rule 4: Level Production Must Be Established

Key Point

Production leveling or load smoothing eliminates variations in flow at different processes and helps maintain stable, smooth production of small lots. How to do this will be discussed in the next chapter. Production leveling is the way all processes can maintain equipment and workers who are ready to produce at the time and in the quantity needed, without carrying excess capacity or excess inventory at each process. This rule also allows you to adapt to small fluctuations in demand by fine-tuning production as conditions change.

Rule 5: Kanbans Always Accompany the Parts Themselves

Key Point

Kanbans are identification tags that certify the need for parts and ensure visual control. This was listed under Rule 1 but is considered a rule of its own by many because the system cannot function if kanbans stray from their parts.

Rule 6: The Number of Kanbans Is Decreased Gradually Over Time

Key Point

Minimize the number of kanbans in order to discover needs for improvement. Line-stopping problems, missing items, and other problems will become visible as you gradually decrease the number of kanbans in circulation. By reducing the amount of stock in the production system, kanban actively sets improvement activities in motion. It will be impossible to neglect improvement by doing this, and so, once again, a rule of kanban is also a critical aspect of lean production.

TAKE FIVE

Take five minutes to think about these questions and to write down your answers:

1. How might the six kanban rules change the way you do work?
2. How would the first two rules work in your area? Draw a picture.

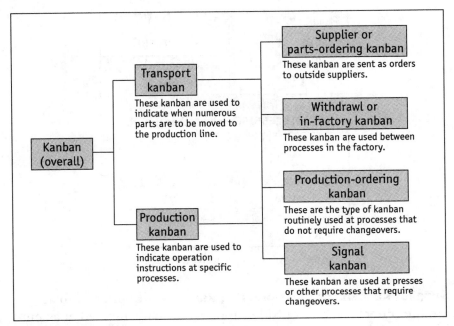

Figure 2-5. Types of Kanban

Types of Kanban

Kanban is the name given to the production order cards in a pull system. Since the Japanese word "kanban" translates as card or sign, and even billboard, you may think that the word refers to visual displays and the visual indicators created in 5S to identify addresses for the placement of tools and parts. But this is not so. Kanbans are the "signs" attached to in-process inventory to indicate production orders. There are many kinds of kanbans that will be discussed next.

Transport Kanbans

Key Point

The first major type of kanban is a transport kanban, which indicates when numerous parts are to be moved to the production line, or between processes in production and assembly. In addition to identifying the part and quantity, transport kanbans indicate where the part comes from and where it is going. There are two basic types of transport kanbans: supplier kanbans and withdrawal kanbans.

From	To
Supplier	**Vision cell**
Shipping post **L5**	Receiving post **M4**
Part no. **760001B245515F Polyurethane. 90D**	
▌║▌█║▌║	Storage location **M-4-B**
Container type **Gaylord**	Number of kanbans **2/3**
Container capacity **1000**	

Figure 2-6. A Typical Transport Kanban

Supplier kanbans or parts-ordering kanbans: Supplier kanbans are the orders given to outside suppliers for parts needed at assembly lines. If the kanban system has extended to the supplier network, then suppliers will deliver on demand as supplier kanbans are received from the factory.

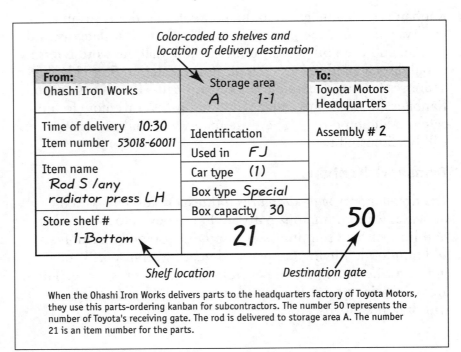

Figure 2-7. Supplier or Parts-Ordering Kanban for Subcontractors

Withdrawal or in-factory kanbans: Assembly lines also use parts and subassemblies that are produced within the factory. These are the kanbans used between processes in the factory; they provide the details needed to withdraw parts from an upstream process.

Withdrawal kanbans are used in many forms depending on the need and type of part being withdrawn—one kanban for a single part or for a container of parts, or a series of kanbans for when parts must be supplied in a certain order for downstream assembly sequences. These may also take the form of a kanban box, as in Figure 2-8, or as a kanban cart for ease of transportation to the downstream process. An example of a kanban cart is on the cover of this book.

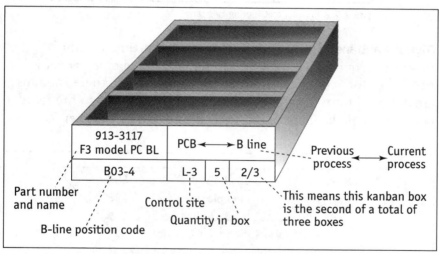

Figure 2-8. A Withdrawal Kanban Box

Production Kanbans

Key Point

The second primary type of kanban is known as a production kanban. Production kanbans indicate operation instructions for specific processes. There are two basic types of production kanbans: production-ordering kanbans and signal kanbans.

Production-ordering kanbans: These are the type that most people think of when they talk about the kanban system. They are routinely used at processes that do not require changeovers. A production-ordering kanban most closely resembles the standard production order used in a push system—it identifies what is to be produced and in what quantity. When a withdrawal kanban authorizes the removal of parts from a line or cell, production-ordering kanbans initiate production to replace the parts that have been removed.

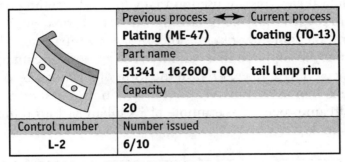

Previous process ◄—► Current process	
Plating (ME-47)	**Coating (TO-13)**
Part name	
51341 - 162600 - 00	**tail lamp rim**
Capacity	
20	
Control number	Number issued
L-2	**6/10**

Figure 2-9. A Production-Ordering Kanban

Signal kanbans: Signal kanbans are used at presses or other processes requiring changeovers to signal when a changeover is needed in the sequence of production kanbans. Triangle kanbans are a special form of signal kanban that call attention to the reorder point. Triangle kanbans are described in detail in Chapter 5.

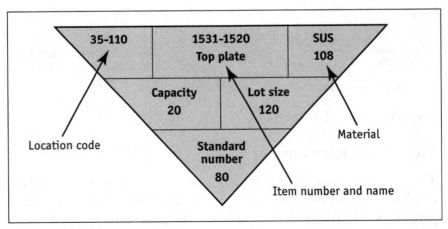

Figure 2-10. A Signal Kanban

Kanbans are usually put in plastic sleeves to keep them clean and undamaged so that they can be used repeatedly. The next chapters discuss how many kanban are needed, how to establish level production and line balancing, and how to circulate kanbans.

TAKE FIVE

Take five minutes to think about these questions and to write down your answers:

1. What would a withdrawal kanban for parts used in your operation look like? Make a sample.

2. What would a production kanban for your operation look like? Make a sample.

In Conclusion

SUMMARY

The kanban system is based on an inventory management system called reordering point method. Though the kanban system grew out of the reordering point system and shares some of its characteristics, kanban is a great improvement over the earlier system in many ways.

The 'push" production system depends on *production work orders* to identify the type and quantity of production to be done at each process. Production work orders are generally used when upstream processes determine how and when goods are moved downstream and how they are controlled between processes. In the kanban system, *kanbans serve as the production order for the pull system.* They follow the goods and indicate what is to be withdrawn from the upstream process. *Order information in a pull system—the kanban—travels upstream from sales to assembly to procurement instead of downstream from planning and procurement to assembly to sales.* Since the downstream processes drive a pull system, originating with the customer order, there is great flexibility in relation to demand, and in-process waste also can be greatly reduced.

There are two primary functions of the kanban system: to serve as the autonomic nervous system for just-in-time production and to improve and strengthen the factory. Kanban is the communication system for lean production. It signals upstream processes when and what to produce and alerts them when problems or changes occur so that they can stop producing. Kanbans serve as work orders; they are *automatic directional devices* that signal standard operations to be engaged at any time based on actual conditions existing in the workplace. They also prevent unnecessary paperwork in start-up operations. Since production only occurs when signaled by a downstream process, in-process inventory and transportation are kept at a minimum and overproduction does not occur.

Kanbans remain attached to the goods that they define and consequently serve as visual controls. Inventory hides problems. Too many kanbans indicate excess in-process inventory. By reducing the number of kanbans, problem areas will come out of hiding so that they can be improved. The kanban system in this way becomes a valuable means to drive out waste and continually improve the production system.

Kanban is the name given the production order cards in a pull system. *The two major types of kanbans are transport kanbans and production kanbans.* Transport kanbans indicate when numerous parts are to be moved to the production line or between processes in production and assembly. In addition to identifying the part and quantity, they indicate where the part comes from and where it is going. There are two basic types of transport kanbans: supplier kanbans and withdrawal kanbans. Production kanbans indicate operation instructions for specific processes. There are two basic types of production kanbans: production-ordering kanbans and signal kanbans.

REFLECTIONS

Now that you have completed this chapter, take five minutes to think about these questions and to write down your answers:

- What did you learn from reading this chapter that stands out as particularly useful or interesting?

- Do you have any questions about the topics presented in this chapter? If so, what are they?

- What additional information do you need to fully understand the ideas presented in this chapter?

Chapter 3

Phase One:
Scheduling Kanban

In Chapter 2 you learned about the functions and rules of the kanban system and reviewed the most important types of kanbans used in a pull system. In this chapter you will learn how to calculate the number of kanbans you need, and how to determine when to produce, what to produce, and by whom each day. This is the scheduling phase of kanban. The goal is maximum flexibility to respond to downstream changes without overproducing.

How Many Kanbans Do You Need?

The kanban system supports level production. It helps to maintain stable and efficient operations. The question of how many kanbans to use is a basic issue in running a kanban system. If your factory makes products using mostly standard, repeated operations, the number of kanbans can be determined using the formula in Figure 3-1.

$$\text{Number of kanban} = \frac{\text{Daily output X (lead-time + safety margin)}}{\text{Pallet capacity}}$$

- Daily output = $\dfrac{\text{Monthly output}}{\text{Workdays in month}}$

- Lead-time = Manufacturing lead-time (processing time + retention time) + lead-time for kanban retrieval

- Safety margin: Zero days or as few days as possible

- Pallet capacity: Try to keep pallet contents small and instead increase the number of deliveries

---

Figure 3-1. How Many Kanban Do You Need?

Key Point

As you can see, *the number of kanbans you need is dependent on the number of pallets or containers and their capacity. Lead times, safety margins or buffer inventory, and transportation time for kanban retrieval are also important factors.*

Several questions must be answered when deciding the number of kanban to use:

1. How many products can be carried on a pallet?

2. How many transport lots are needed, given the frequency of transport?

3. Will a single product or mixed products be transported?

The relationship between the order-to-delivery period and the production cycle is also relevant. You will want to consider the following questions:

1. *Order-to-delivery time.* What quantity do assembly processes require and in what amount of time?

2. *Production cycle or lead time.* Consider the following:
   a. The time it takes to send withdrawal kanban to the previous process after removing the kanban.
   b. The time that elapses until processing is begun after exchanging withdrawal kanban for WIP kanban.
   c. The time it takes to produce supply lots.
   d. The time it takes to store the lot to be processed.
   e. The time it takes to transport processed items to the assembly line.

Key Point

In lean production, how to determine the number of kanbans used is less important than how to improve the production system so that this number can be reduced. To achieve a minimum number of kanbans several important improvements must occur:

1. Production must be done in small lots:
   a. Reduce setup times to a minimum.
   b. Cut lead times to a minimum.

2. Buffer stocks that are kept as safety margins against market fluctuations and production instability must eventually be eliminated.

Short setup times make it possible to respond quickly to change. A short production cycle allows you to reduce the number of kanbans to the minimum since reliable information about changes is easily accessible and the system responds rapidly. Kanban reduction as an improvement method is explored in more detail in Chapter 5.

If your factory produces custom-order products kanbans can still be used. One kanban needs to be made for each order. It must include the work order information and when to produce the item ordered. If finished products are to be delivered to different downstream processes, the next site should also be indicated on the kanban, as well as when it will be withdrawn by the next process. Thus, even for custom-order shops kanban functions as a visual control for the production process.

Takt Time

Key Term

Takt time is an important element to understand in production scheduling. *Takt time is the rate at which products or parts must be produced to fulfill customer orders.*

Example

For example, if customer demand is 240 products a day and production time equals 480 minutes per day, then the takt time — the time you have to produce one product—is 2 minutes. You may be able to produce the product in less than that and therefore could produce more than 240 per day, but you will restrain from doing more than what is needed, thus avoiding overproduction. *Takt time, therefore, is not a measure of what you are capable of but a calculated number designed to match production to market demand.*

The formula for calculating takt time is:

$$\text{Takt time} = \frac{\text{operating hours per day}}{\text{output demand per day}}$$

$$\text{Output demand per day} = \frac{\text{output demand per month}}{\text{operating days per month}}$$

Example

If 20,000 units must be produced in a month and there are 20 working days in a month then 1000 units must be made each day. If there are 480 minutes of operation in an 8-hour workday then each unit must be produced in .48 minutes.

480 minutes/1000 units = .48 minutes

Takt time sets the rhythm or pace of production. Like clapping your hands in time to a piece of music, takt time claps the beat of the market. When orders are up, takt time will set a faster beat; when orders are down, takt time will be slower. *The kanban system is the means used to regulate the rhythm of production to takt time.* Load leveling discussed later in this chapter is the means used to keep the rhythm steady.

Key Point

TAKE FIVE

Take five minutes to think about these questions and to write down your answers:

1. What would the takt time be for producing 40,000 units per month? Do the calculation.
2. How many units do you produce or operations do you complete in a day in your line or cell?
3. How long does it take your team to make one unit or complete one operation?

How Many Operators Are Needed?

Line Balancing

Key Term

Line balancing is the process by which work is evenly distributed to workers to meet takt time. Some operations take longer than others, which may cause one or more operators to wait for the next part; and some operations need more than one operator. Line balancing helps to use every worker well and makes sure that no one is idle for too long or working too much. Several steps are required to balance the line.

How-to Steps

New Tools

Step one: To balance the line you must understand the current conditions. Draw a map of the process identifying each operation and the number of operators currently on the line. Note the cycle times of each operation and add them together to get the total process cycle time. Figure 3-2 shows an example of a *process map* of the operations of one cell. You can also use a *table showing current state data*, like the one in Figure 3-3. Note the measurements for the current conditions in each operation and for each operator.

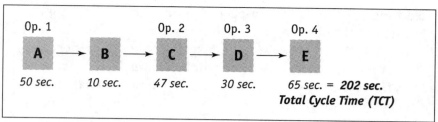

Figure 3-2. Process Map for Line Balancing

	Machine	Deburr	Crimp	Test	Mark
Cycle time	50 sec.	10 sec.	47 sec.	30 sec.	65 sec.
Changeover	60 min.	0	5 min.	5 min.	5 min.
Operators	1	0	1	1	1
Uptime	87%	100%	99%	99%	99%
Availability	27,600 sec.	27,600 sec.	27,600 sec.	27,600 sec.	27,600 sec.

Figure 3-3. Table of Current State Data

Step two: The next step is to create an *operator balance chart* to give you a visual representation of the data you collected. An operator balance chart compares the cycle times of each operation to the takt time. It will help you understand the current condition and can also be used to plot the desired condition or balanced line in relation to takt time. It is a visual display of individual work operations, operation times as they compare to total cycle time and takt time, and operators at each position. An operator balance chart will show you where opportunities for improvement exist. In the example in Figure 3-4 you can easily see that the line is out of balance in the Current State.

Step three: Next determine the number of operators needed. The formula is:

$$\# \text{ of operators needed} = \frac{\text{Total cycle time}}{\text{Takt time}}$$

For example: $\dfrac{202 \text{ total cycle time}}{60 \text{ takt time}} = 3.36$

34

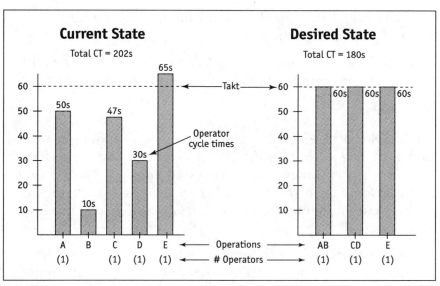

Figure 3-4. Operator Balance Chart

In the example shown, the number of operators needed is 3.36, in other words not quite enough to keep four operators working full time, though there is more work than three can do.

This shows you that you have an opportunity to redesign the line to better use the people assigned to it and to better adhere to takt time. The team can work on reducing waste and improving the cycle time of each operation so that three people will be sufficient to handle the five operations. The fourth person now working part time on this line can then be used on a line that is deficient in workers. Whenever you do the formula to determine how many workers are needed and the result includes a decimal amount less than or equal to 0.5 (in our example the fraction is 0.36), it is a good indication that reducing the line by one person is a realistic goal.

Example

In our example, the goal is to have three workers create the part within takt time — each worker spending no more than 60 seconds for their operations. This means they would complete the five operations in 180 seconds, a reduction of 22 seconds in the total cycle time. One solution to reaching this goal is shown in the Desired State in Figure 3-4. It proposes that the first four operations be combined into two — A+B and C+D. Operations C, D, and E will have to be simplified to be able to complete them in 60 seconds or less. This is the target. As you can see, line balancing will support your takt time, which, in turn, impacts your kanban effort.

The challenge is to find a way for this to be achieved. The solution lies in the concept of *full work through standardization.*

Full Work

Kanban demands efficient and timely use of machines, materials, processes, *and workers.* In Figure 3-4 the ideal condition is for all three workers to achieve the condition of *full work—each operator safely performing operations at takt time without any waiting* or, in this example, at 60 seconds per operation. How will this be accomplished?

In the Current State, operator E is overworked and probably always seems to be behind. The other operators have idle time, because the first four operations are being performed at less than takt time. You can see that once the decision is made to shift to a pull system, full work will quickly become the essential target for every line to achieve in order to fulfill the demands of takt time without creating excessive wait time or overwork. Finding ways to evenly distribute the work among the operators on each line and within each cell is the task of the teams in each process.

In Figure 3-5 you can see two ways to redistribute work and eliminate the need for one worker. Only one of these solutions results in *full work.* The difference is that the good example organizes the work in relation to takt time; the poor example simply divides the total cycle time by 4 and distributes it equally among the four workers. This result is that all of them are under worked. If, on the other hand, you redistribute with full work as the goal, then further opportunity to streamline operations and improve efficiency is revealed. The fourth person is only working 40 percent of the time. Standardization and waste reduction may be able to reduce the need for this person for this activity as well.

Please note that *improvements toward full work are not aimed at reducing the workforce but at reducing waste and synchronizing all processes to customer demand.* Removing people from one line means that they will be placed on another line where more workers are needed. Redistribution plans toward full work will also reveal the cells that are overworked in relation to takt time and need more workers to meet demand. Also, as takt time

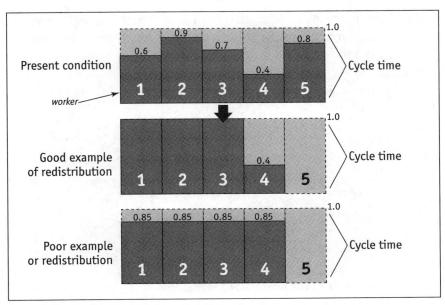

Figure 3-5. Line Balancing to Achieve Full Work

increases in periods of increased market demand, workers will need to be moved to the cells and lines where the demand is greatest. If full work is not being achieved there will be no possibility to respond to these changes in demand, and kanban and the pull system will fail.

Standardization: Once the full work goals are established, then improving visual controls, eliminating waste, reducing setup times, and gradually reducing kanbans in the system can reduce individual cycle times. Some ways to improve within the kanban system are discussed in Chapter 5.

Key Term

Standardization is also an important way to achieve full work. It is explained in detail in another book in the Shopfloor Series. *Standardized work consists of the work procedures that everyone agrees are the best method and sequence for each process.* At some point standardization will have to be implemented to improve cycle times so that the number of kanbans being used in the factory can be reduced and WIP inventories can be brought to a minimum.

Automated machines: The term "full work" was originally applied to automated machines. In this respect it refers to the need to set limits on machine production so that overproduction does not occur. Limit switches are installed to restrain machines in the preceding process from producing until they are signaled to do so. Figure 3-6 is an example of how this works.

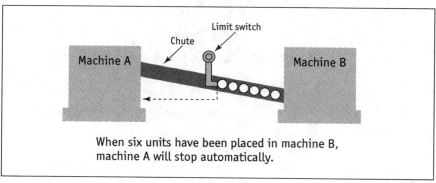

When six units have been placed in machine B, machine A will stop automatically.

Figure 3-6. Full Work System for Machines

New Tool

A line balancing worksheet: A worksheet such as the one provided on pages 39 to 41 will take you through the steps of balancing a line to help achieve full work.

TAKE FIVE

Take five minutes to think about these questions and to write down your answers:

1. What is the cycle time of each operation in your cell or line?
2. Make an operator balance chart showing the cycle times of each operator at each station or operation. What does it show?
3. Do you know the takt time for your process?
4. What could you do to improve the balance of work?

Line Balancing Worksheet

Purpose: To make decisions necessary for adding continuous flow elements to the future-state map.

Directions: 1. Discuss each item in sequence.
2. Make sure the team comes to a decision for each item.
3. Record the "minority" opinions; they may prove useful later.
4. Ensure that the team scribe records the decisions.

Takt Time

Your takt time is   _____

Your pitch is   _____

Operator Balance Chart

1. Review your current-state attributes.

Operations ⟶							
Cycle Time							
Changeover							
Operators							
Uptime							
Availability							

Line Balancing Worksheet

2. Create a current-state bar chart.

Operator Balance Chart–Current State

3. Determine the number of operators needed to meet takt time.

DETERMINING THE NUMBER OF OPERATORS

$$\text{\# of operators} = \frac{\text{Total cycle time}}{\text{Takt time}}$$

$$\text{\# of operators} = \frac{\underline{\hspace{1cm}} \text{ (Total cycle time)}}{\underline{\hspace{1cm}} \text{ (Takt time)}} = \underline{\hspace{0.5cm}} \text{ Operators}$$

Line Balancing Worksheet

4. After going to the shopfloor and studying the current state further, set a new total cycle time (if appro-priate) and decide on the number of operators needed.

$$\text{\# of operators} = \frac{\underline{\quad\quad} \text{ (Total cycle time)}}{\underline{\quad\quad} \text{ (Takt time)}} = \underline{\quad} \text{ Operators}$$

5. Complete a proposed operator balance chart by creating a bar chart of the future state.

Operator Balance Chart–Current and Proposed

Load Leveling or Smoothing—*Heijunka*

Load leveling is the equalization of quantities and types of products on any line. It is also called production smoothing or *heijunka* and is an important part of lean manufacturing. Load leveling puts customer orders in a sequence that smoothes day-to-day variations while meeting longer-term demands. It eliminates peaks and valleys in the workload as well as excess production.

For instance, if in one week customers order batches of 250 of Product A, 500 of Product B, and 250 of Product C, load leveling would call for a sequence on the line as follows: A, B, C, B, A, B, C, B ... and so on.

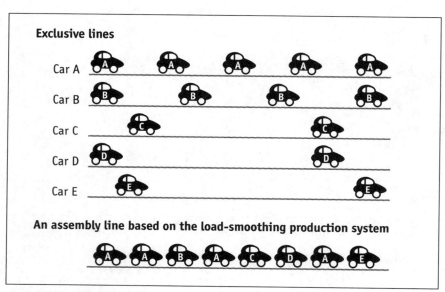

Figure 3-7. Load Smoothing Auto Production

If load leveling is part of the production plan it also results in increased flexibility in the plan; it makes it easier to change production orders in midstream in response to changes in downstream conditions.

If a line produces 150 units each day and a change increases the demand to 155 units, you will probably take that change in stride without thinking about changing the system or your production capacity. If, however, the change requires 200 units be produced it will be difficult for you to respond to this change without going into overtime, new equipment, or new personnel. Yet if this change is responded to gradually by adding the 50 additional units over a month with load leveling, the increased demand can be handled easily.

If it is known that demand will increase a certain percentage over the next month, daily output can be increased five to eight units to respond effectively and seamlessly to the change. If the rules for changing the production plan require that changes can only be made at certain times—monthly meetings perhaps—or if approval of plans takes too much time, then early response to upcoming fluctuations cannot be responded to adequately with gradual load smoothing. The kanban system will fail if there is no load-smoothing system of production.

TAKE FIVE

Take five minutes to think about these questions and to write down your answers:

1. Does the number of units you produce change significantly during the month?
2. Do you produce more than one product type?
3. How large are the lots that you produce for each product type?

Level Production Compared to Shish-Kabob Production

BACKGROUND INFO

There are many ways to create production schedules. A push system based on batch or large-lot production methods creates schedules for once-a-month or once-a-week production. A once-a-month production schedule takes the estimated demand for a month for each product and produces that amount all at once each month. Once-a-week production divides the monthly projection into weekly amounts and produces those batches each week. When markets were more stable and product variety smaller these product-out approaches were adequate, though they do not focus on customer needs and waste remains hidden throughout the manufacturing process.

Many factories are turning to daily production schedules and creating an integrated production line using a "shish-kabob" approach. The daily schedule takes the monthly projection and divides it into daily quotas to meet the monthly figure. Though

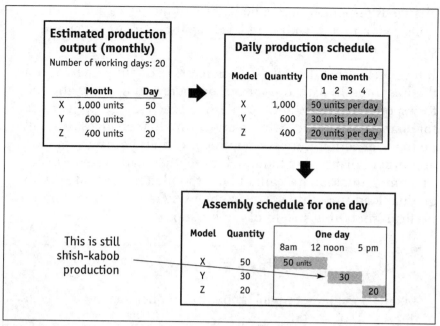

Figure 3-8. Shish-Kabob Production (Once-a-Day Scheduling)

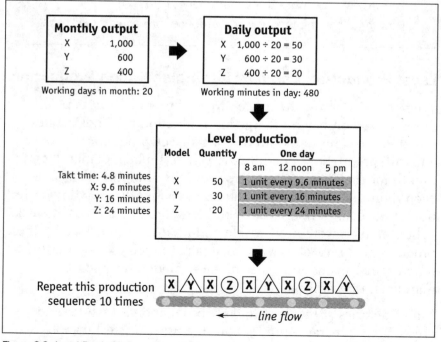

Figure 3-9. Level Production

there are mixed-model production lines, they are still being done in batch production style—each part of the day being used for production of a different model or part. Setup is kept at a minimum each day and seems to be the controlling factor in how the day is scheduled. This is not true heijunka.

In level production, setup times have been reduced to such a point that setup is no longer a scheduling factor. Instead, production of different models is evenly spread out in order to match the current sales trends. Level production is entirely a "market-in" approach.

In Figure 3-10 the differences between the shish-kabob production approach and level production are summarized.

A pull system is based on actual need, not projections of customer demand. Succeeding with a pull system depends on scheduling production based on the needs of the downstream operation closest to the customer. Pull and continuous flow methods must initiate and signal all other activities using kanbans. There must be zero downtime, little or no changeover time, and great worker flexibility.

	Shish-Kabob Production	Level Production
Production Philosophy	**Product-out (production-centered) philosophy.** "Produce just what is easy to make, just when it is easy to make it, and in just the amounts that are easy to make."	**Market-in (market-centered) philosophy.** "Produce just what is needed, just when it is needed, and in just the amounts needed."
Production Method	**Lot (shish-kabob) arrangement.** Arrange products into large model-specific lots to minimize changeovers.	**Cycle time arrangement.** Arrange products into assortments that match market needs and can be manufactured within the cycle time in an in-line production configuration.
Approach to Efficiency	**Emphasis on individual process efficiency.** The production pitch is based on the rhythm of individual processes with maximum efficiency sought at each process.	**Emphasis on overall line efficiency.** We try to improve the efficiency of the entire line within the framework of the cycle time.
Approach to Machines	**High-speed, general purpose, large, and expensive machines.** We need faster machines to handle large lot volumes, which usually means we need a large, expensive, general purpose machine.	**Moderate-speed, specialized, small, and inexpensive machines.** Our machines need only be fast enough to keep up the cycle time. The important thing is that the machines be small and specialized enough to fit right into the production line to handle one-piece flow operations. Such machines are usually much less expensive than large, general purpose machines.
Inventory and Lead-Time	**Large inventories and long lead times.** When workpieces are worked on in lots, retention is inescapable. Retention accumulates in-process inventory and results in longer lead times and a greater need for conveyance.	**Small inventories and short lead times.** When workpieces flow along one piece at a time within the cycle time, there is very little in-process inventory, which means shorter lead times and almost no need for conveyance.

Figure 3-10. Differences Between Shish-Kabob Production and Level Production

TAKE FIVE

Take five minutes to think about these questions and to write down your answers:

1. What kind of production scheduling is done at your plant?
2. What are some of the advantages of load leveling?

In Conclusion

SUMMARY

The kanban system supports level production. It helps to maintain stable and efficient operations. *The number of kanbans you need is dependent on the number of pallets or containers and their capacity. Lead times, safety margins or buffer inventory, and transportation time for kanban retrieval are also important factors.* However, *in lean production, how to determine the number of kanbans used is less important than how to improve the production system so that this number can be reduced.* To achieve a minimum number of kanbans, production must be done in small lots and you must eliminate buffer stocks that are kept as safety margins against market fluctuations and production instability.

Takt time is an important element to understand in production scheduling. *Takt time is the rate at which products or parts must be produced to fulfill customer orders. Takt time is not a measure of what you are capable of but a calculated number designed to match production to market demand.* Takt time sets the rhythm or pace of production. Like clapping your hands in time to a piece of music, takt time claps the beat of the market. When orders are up, takt time will set a faster beat; when orders are down takt time will be slower. *The kanban system is the means used to regulate the rhythm of production to takt time.* Load leveling is the means used to keep the rhythm steady.

Line balancing is the process by which work is evenly distributed to workers to meet takt time. Some operations take longer than others, which may cause one or more operators to wait for the next part; and some operations need more than one operator. Line balancing helps to use every worker well and makes sure that no one is idle for too long or working too much.

It is possible to redesign the production lines to better adhere to takt time. The team can work on reducing waste and improving the cycle time of each operation. The solution for doing this lies in the concept of *full work through standardization.* The ideal is for all workers to achieve the condition of *full work—*

each operator safely performing operations at takt time without any waiting. Standardization is an important way to achieve full work. *It consists of the work procedures that everyone agrees are the best method and sequence for each process.* At some point, standardization will have to be implemented to improve cycle times so that the number of kanbans being used in the factory can be reduced and WIP inventories can be brought to a minimum.

Load leveling is the equalization of quantities and types of products on any line. It is also called production smoothing or *heijunka* and is an important part of lean manufacturing. Load leveling puts customer orders in a sequence that smoothes day-to-day variation while meeting longer-term demands. It eliminates peaks and valleys in the workload as well as excess production. If load leveling is part of the production plan it also results in increased flexibility in the plan; it makes it easier to change production orders in midstream in response to changes in downstream conditions.

If it is known that demand will increase a certain percentage over the next month, daily output can be increased five to eight units to respond effectively and seamlessly to the change. If the rules for changing the production plan require that changes can only be made at certain times—monthly meetings perhaps—or if approval of plans takes too much time, then early response to upcoming fluctuations cannot be responded to adequately with gradual load smoothing. The kanban system will fail if there is no load-smoothing system of production.

There are many ways to create production schedules. A push system based on batch or large lot production methods creates schedules for once-a-month or once-a-week production. A once-a-month production schedule takes the estimated demand for a month for each product and produces that amount all at once each month. Once-a-week production divides the monthly projection into weekly amounts and produces those batches each week. When markets were more stable and product variety smaller these product-out approaches were adequate, though they do not focus on customer needs and waste remains hidden throughout the manufacturing process. *In level production setup times have been reduced to such a point that setup is no longer a scheduling factor.* Instead, production of different models is evenly spread out in order to match the current sales trends. Level production is entirely a "market-in" approach.

REFLECTIONS

Now that you have completed this chapter, take five minutes to think about these questions and to write down your answers:

- What did you learn from reading this chapter that stands out as particularly useful or interesting?

- Do you have any questions about the topics presented in this chapter? If so, what are they?

- What additional information do you need to fully understand the ideas presented in this chapter?

Chapter 4

Phase Two: Circulating Kanban

Now you are ready to use the kanban system. In this chapter you will learn how to create and circulate kanban cards to put the pull system in motion.

You may remember from Chapters 1 and 2 that one of the most important changes occurring in a pull system using kanban is that operators produce only when signaled to do so by a downstream process. This is the key to the kanban system, so let's talk about that more as you begin to put kanban to work.

When Do You Produce?

Key Point

Remember, the downstream process pulls from the upstream process. And the upstream process only produces to replenish what has been withdrawn. This means that *you only produce more products or parts when you receive a production-ordering kanban*. This will only happen when a downstream process has come to you with a withdrawal kanban authorizing them to remove a part or product you have already produced. They will take the part and leave a production-ordering kanban in its place. A good way to collect production kanbans is in a kanban post, shown in Figure 4-1.

Producing only when signaled is a big change in behavior for any operator who has been trained in a push system. In a push system with batch or large-lot production, workers are rewarded for being able to produce more products in less time. You produce as much

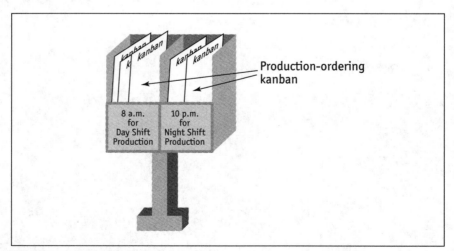

Figure 4-1. Production-Ordering Kanban Post

as you have material available to use. The upstream processes *push* parts and subassemblies through the system as fast as they can and sub-assemblies mount up in the downstream process areas. Typically, the downstream workers are overloaded with production orders to complete against tight schedules.

Key Point

In a pull system with kanban the factory will be synchronized to the customer and workers will be distributed in the process to best meet demand. Production orders come from downstream and you will produce to replenish what has been withdrawn, instead of to get rid of in-process inventories.

What Do You Replenish?

Example

You only replenish what has been withdrawn from your work process. Depending on your area or process, this can be any of the following:

1. Raw material

2. Parts

3. Components or subassemblies

4. Finished goods

If you work in the inventory warehouse for raw materials and parts purchased from outside suppliers or other plants, you will replenish raw materials as they are withdrawn for use in your subassembly production processes. If you work in subassembly, you will

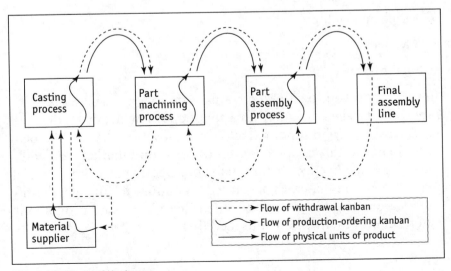

Figure 4-2. Chain of Kanbans

produce the parts and components as they are withdrawn for final assembly. If you are in final assembly, you will produce finished goods as customers order the products.

TAKE FIVE

Take five minutes to think about these questions and to write down your answers:

1. How will kanban change the way you work?
2. How do you know when to produce parts?

When Do You Pull?

You pull material from the upstream process only as you need it to replenish what you have used to produce parts, components, or final assemblies.

Where Can You Pull From?

Example

Places you might pull from include:

1. A supplier:
   a. An external supplier
   b. Other plants within your company

2. An upstream operation

3. A kanban quantity

4. A feeder cell

5. Supermarkets

In relation to your own process, upstream to you may be the preceding process or it may be a supplier of raw materials or subassemblies. The supermarket is an innovation that will be discussed later in this chapter. When using a supermarket, you withdraw materials needed to complete your operations from a special place where the materials are stored. Withdrawal kanbans identify the location of these materials so that there is no confusion about where to find them when you need them.

What Do You Pull?

Example

The material you pull from the upstream process can be contained in a number of ways. It may be any of the following:

1. A single part, a subassembly, or final assembly

2. A container or pallet of parts

3. A production lot

4. A complete customer order

Whatever you pull must have a withdrawal kanban with it, which tells the exact quantity that is contained in the pallet or lot that you withdraw.

TAKE FIVE

Take five minutes to think about these questions and to write down your answers:

1. When and why do you pull parts from the preceding process?
2. Where is your preceding process?

What Belongs on a Kanban Card?

We have shown a few examples of kanban cards throughout this book. As you have seen, they will vary to serve the purpose for which they are intended. Ideally, they will be as thorough as possible and include the following kinds of information:

1. The material, part, subassembly, or assembly number

2. A description with a drawing or photo, if possible

3. The previous process—where did it come from?

4. The next process—where does it go?

5. The internal or external supplying process (origin)

6. The customer or factory order number

7. What, when, and how much to withdraw

8. What, when, and how much to make

How Do You Attach Kanban Cards?

Kanban cards are usually kept in plastic sleeves to protect them from dirt or tearing, which allows them to be read and handled easily and used repeatedly. They are attached to standardized containers that hold the parts to be withdrawn. One card can represent multiple products in one container or in multiple containers to correspond to a single production order.

Example

Container options: There are a variety of creative solutions, including:

1. Pallets
2. Bins
3. Trays
4. Kits

5. Boxes of various sizes
6. Carts
7. Trucks, trailers, railroad cars

TAKE FIVE

Take five minutes to think about these questions and to write down your answers:

1. What would a kanban card for the part you produce look like? Make a sample one.
2. What type of container is used to transport the parts you make to the downstream process?
3. Where are the raw materials or parts stored that you use to produce what you make?

Steps for Circulating Kanbans

Let's look closely at the process of circulating the kanban cards. There are a number of steps involved.

-to Steps

1. When the assembly line uses parts, they place the associated kanbans in the withdrawal kanban post.

2. The carrier removes a withdrawal kanban from the withdrawal kanban post and takes it to the preceding process to replace the items that have been used in assembly.

3. The carrier then removes a production kanban from the pallet or container and puts it in the production kanban post for that process. The withdrawal kanban is placed on the replenished pallet, which is then transported back to the assembly line area.

4. The production kanban removed from the withdrawn pallet serves as a production order to produce the withdrawn items.

5. Empty pallets are placed in the designated space.

6. As units are produced they are placed with the production kanbans in a storage area located within or near the production area so that carriers from subsequent processes can withdraw them at any time.

7. Withdrawal kanbans are taken to the preceding process to replace parts or subassemblies used to produce the replacement items in the same manner as in step 2.

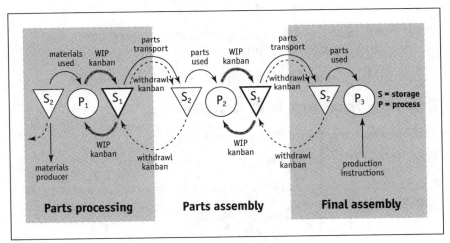

Figure 4-3. How Kanbans Are Circulated

In this simple way, the steps above create a chain reaction that is set in motion all the way back to outside suppliers. Only final assembly needs to be notified of changes in plans. Through the kanban system any change will automatically be communicated to upstream processes in the natural order of pull production.

Rules for Circulating Kanbans

Make sure the following rules are adhered to:

1. Each pallet or container has one kanban sheet or card.

2. Kanban cards always accompany the material.

3. The quantity indicated on the card is the same as the quantity in the container.

4. Kanban posts hold kanbans for material that is withdrawn or produced and indicate material that has been used in production.

5. When production begins in a subsequent process, withdrawal kanbans are put in withdrawal kanban posts to signal the need for replacement of materials from the upstream process.

6. Production kanbans are placed in production kanban posts in the same sequence the material is withdrawn.

7. Production occurs in the same order that the production kanban collect in the production kanban post.

An example of a production kanban post was seen in Figure 4-1.

Rules for Transport with Kanbans

These rules will ensure proper transport:

1. Deliveries take place several times each day, usually hourly.

2. Empty pallets or containers are placed in the designated area of the preceding process.

3. Physical delivery points are clearly indicated.

4. The space dedicated for delivered items is small enough to avoid accumulation of excess stock and large enough to hold an average delivery amount—and certainly no more than needed to meet the typical material needs in one day of production.

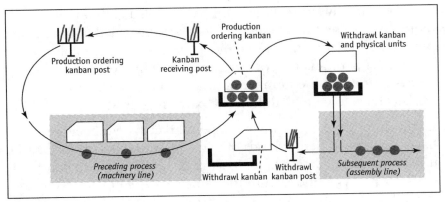

Figure 4-4. Transport with Kanban Circulation

TAKE FIVE

Take five minutes to think about these questions and to write down your answers:

1. How does the withdrawal kanban replace a production kanban? Where does the production kanban go?

2. What are the special rules of transport that need to be adhered to?

3. Do you have an area that could be created for downstream processes to leave empty pallets?

4. How do you know in what sequence to produce new parts?

Supermarkets

In Chapter 1 we mentioned that the kanban system was inspired by the way supermarkets work to supply customers with products. The kanban system relies on a replenishment system similar to the supermarket. When the downstream process (the customer) withdraws parts from an upstream process (the supermarket) the upstream process replenishes the withdrawn product by making another amount equal to what was withdrawn.

Key Term

The areas where produced parts, components, subassemblies, and finished goods are stored are called supermarkets in the kanban system. These *supermarkets* or stores are located near the area where the products or parts are produced. Customers — the next process — come to the area to retrieve the parts they need. Withdrawal and production kanbans control this process, insuring that there are always parts available when needed, but that no more than needed will be made. Subassembly *kits* that have been preselected and organized to reduce assembly time can also be stored in supermarkets between major steps in the assembly process, reducing the number of places that downstream process workers have to go to withdraw the parts they need.

If you cannot use an in-process supermarket system because you have too many different kinds of parts, then use the FIFO (first in, first out) system. Create lanes to store dissimilar parts so that the parts are always withdrawn in the same sequence they were made; the oldest (first in) is the first used (first out).

Improvement in the process will be aimed at reducing the number of kanban needed and therefore the size of the *supermarkets* or WIP inventories that must be maintained. Ultimately zero kanban is the *ideal* goal, eliminating the need for WIP storage. However, the supermarket can be a helpful way of thinking about how the transfer of goods occurs down the line. It is a way of controlling inventory and keeping it at a minimum, and controlling production without having to schedule every supplying process. See Figure 4-5.

Figure 4-5. Circulation And Transport Using Kanban with Supermarkets

Water Beetles and Milk Runs

Key Term

The term *water beetle* comes from the insect called a whirligig. It moves swiftly across the surface of water, spins, and makes rapid and unexpected changes in direction. In the Toyota Production System this name was given to the conveyor—*an operator who delivers parts to the other operators in his cell or on the line so that they don't have to get up to replenish their work stations to make the next set of products.*

In the kanban system frequent transportation of parts becomes necessary. Usually hourly deliveries are needed. For simplicity, we have described the withdrawal and production process in terms of a single withdrawal location. The reality is that most production items to be produced require numerous parts that come from several preceding processes. This can be a time consuming and confusing aspect.

Key Point

Designating a carrier to withdraw parts from the preceding processes and deliver them to workers greatly simplifies the kanban process. Also, this person quickly becomes an expert in the withdrawal and production kanban process, making it possible to identify and eliminate errors.

The benefits of the water beetle: When a water beetle supports the line:

1. Operators don't have to leave the work area to find parts or tools.

2. Waiting time becomes visible.

3. The *water beetle* becomes the pace maker for the line; delays and progress become visible.

4. *The water beetle* can fill-in for absent operators or for less than "full work" operations.

The tools needed by the water beetle: To do the job well, the water beetle will need:

1. A pushcart with:
   a. Small wheels for easy maneuvering
   b. A checklist for pickups and deliveries to be made
   c. A layout of cells or lines so that pick up and placement can occur easily

2. A picking list to:
   a. Clearly identify *what* and where to pick up
   b. Clearly identify where and *when* to pick up
   c. Limit the number of items on the list

3. Visual displays to help identify supplies to be picked up or delivered

4. Pickup buckets for ease of transfer to and from the pushcart

5. Storage shelves where:
   a. Items are stored so that they are easily transferred to the pushcart—nothing above the shoulder or below the knee
   b. Gravity is used whenever possible to minimize lifting

6. A supply area at the line that:
   a. Makes transfer from pushcart to the line easy
   b. Is large enough for only one *kit*

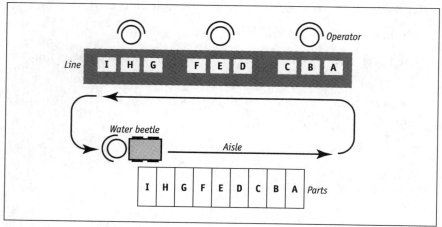

Figure 4-6. The Water Beetle at Work

Key Term

Milk runs: *Milk runs are the name given to the path the water beetle takes in his or her delivery and pickup runs.* It shows a layout that allows the water beetle to serve a number of cells in one pass from supermarket and back again.

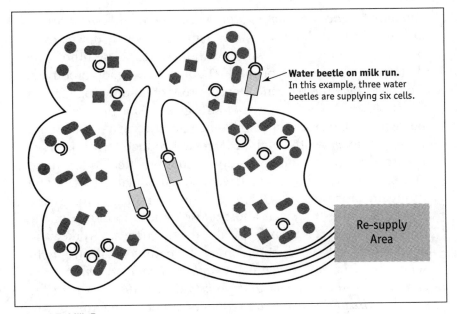

Water beetle on milk run.
In this example, three water beetles are supplying six cells.

Re-supply Area

Figure 4-7. Milk Runs

TAKE FIVE

Take five minutes to think about these questions and to write down your answers:

1. Would it help to have a designated carrier be in charge of transport for your cell or line?

2. Is your plant laid out so that a milk run could be established effectively? What would you need to do to make this possible?

In Conclusion

SUMMARY

You only produce products or parts when you receive a production-ordering kanban. This will only happen when a downstream process has come to you with a withdrawal kanban authorizing them to remove a part or product you have already produced. They will take the part and leave a production-ordering kanban in its place. Producing only when signaled is a big change in behavior for any operator who has been trained in a push system. In a *push system* with batch or large-lot production, workers are rewarded for being able to produce more products in less time. You produce as much as you have material available to use. The upstream processes *push* parts and subassemblies through the system as fast as they can and subassemblies mount up in the downstream process areas. Typically, the downstream workers are overloaded with production orders to complete against tight schedules. *In a pull system with kanban the factory will be synchronized to the customer and workers will be distributed in the process to best meet demand.* Production orders come from downstream and you will produce to replenish what has been withdrawn, instead of to get rid of in-process inventories.

You only replenish what has been withdrawn from your work process. This can be any of the following depending on your area or process: raw material, parts, components or subassemblies, or finished goods. You pull material from the upstream process only as you need it to replenish what you have used to produce parts, components, or final assemblies. You can pull from a supplier—external or other plants within your company, an upstream operation, a kanban quantity, a feeder cell, or in-process *supermarkets*. Withdrawal kanbans identify the location of these materials so that there is no confusion about where to find them when you need them. The material you pull from the upstream process can be contained in a number of ways—a single part, subassembly, or final assembly; a container or pallet of parts; a production lot; or a complete customer order. Whatever you pull must have a withdrawal kanban with it, which states the exact quantity that is contained in the pallet or lot that you withdraw.

Improvement in the process will be aimed at reducing the number of kanban needed and therefore the size of the supermarkets or WIP inventories that must be maintained. The supermarket can be a helpful way of thinking about how the transfer of goods occurs down the line. It is a way of controlling inventory and keeping it at a minimum, and controlling production without having to schedule every supplying process.

The term *water beetle* comes from the insect called a whirligig. It moves swiftly across the surface of water, spins, and makes rapid and unexpected changes in direction. In the Toyota Production System this name was given to the conveyor—*an operator who delivers parts to the other operators in his cell or on the line so that they don't have to get up to replenish their work stations to make the next set of products.*

In the kanban system frequent transportation of parts becomes necessary. Usually hourly deliveries are needed. For simplicity, we have described the withdrawal and production process in terms of a single withdrawal location for simplicity. The reality is that most production items to be produced require numerous parts that come from several preceding processes. This can be a time consuming and confusing aspect. *Milk runs are the name given to the path the water beetle takes in his or her delivery and pickup runs.* It shows a layout that allows the water beetle or conveyor to serve a number of cells in one pass from supermarket and back again.

REFLECTIONS

Now that you have completed this chapter, take five minutes to think about these questions and to write down your answers:

- What did you learn from reading this chapter that stands out as particularly useful or interesting?

- Do you have any questions about the topics presented in this chapter? If so, what are they?

- What additional information do you need to fully understand the ideas presented in this chapter?

Chapter 5

Phase Three:
Improving with Kanban

Fine-Tuning Production by Reducing Kanbans

As we discussed in Chapter 2, one of the functions of the kanban system is to improve and strengthen the factory as a whole. How does it do this?

Kanban Highlights Abnormalities

Key Point

The kanban system reveals waste and problems very quickly. If setup times are too slow this will become evident immediately when you try to work to takt time. Using kanbans makes it easier to spot any tie-ups in the process due to setup time. Any process that is overproducing will have no room for the inventory they have produced because the pallet, bin, shelf, or supermarket designated for holding the material will be overflowing. The way you design these areas so that this cannot happen is discussed below in the section on visual controls. Underproduction will also be evident by empty shelves, etc. Many kinds of waste and problems are revealed through kanban, thus enabling production to be fine-tuned.

Figure 5-1. Fine-Tuning Production

Reducing Kanbans Reduces Stock

Key Point

Reducing kanbans means reducing stock, which ultimately elimi-nates the safety stocks usually maintained as insurance against production fluctuations. Supervisors are given freedom to start with as many kanbans as they want and then, as the process con-tinues, to reduce the number of kanbans one by one until the minimum possible is reached. In this way too, the kanban system becomes a method of fine-tuning the production process. Figure 5-2 shows this function of the kanban system, comparing stock levels to the water levels in a pond. As the water (stock) level lowers, problems (X) are exposed and can be addressed.

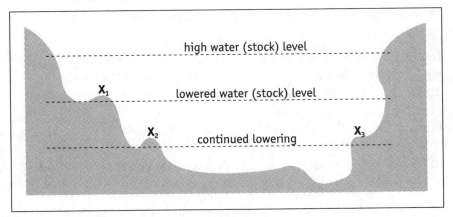

Figure 5-2. Improvement Function of the Kanban System

When you begin to implement the kanban system you start with the number of kanbans equal to current production flows. How to determine the number to use was described in Chapter 3 where you learned that the number of kanbans changes automatically based on customer demand. It's important to note here that reducing kanbans to stimulate improvement activities is a separate process from changing the number of kanbans in response to cus-tomer demand. Changing kanbans in response to customer demand is related to maintaining takt time as orders increase or decrease. These changes will not be dramatic and will be made as part of the scheduling process. It allows production to be com-pletely flexible.

Reducing the total number of kanbans in order to identify problem areas and reduce stock is different. Supervisors consciously reduce kanbans one by one to discover what the minimum number of kanbans can be in each production area.

This is a trial and error process and should be worked on as a team. Supervisors must pay attention to when the number of kanbans is too low. If the safety-stock level is too small, you will lose your motivation to improve. However, if the level is too high, work will become boring. What you want to find is the level that tightens up your process without causing you to feel constricted; rather it should begin to feel good to work at the pace you have set. As you use kanban reduction to improve your process your work should become more enjoyable. This is a critical element to pay attention to as you improve.

If you remove just one kanban, perhaps you will notice only a slight change. Your team may be working just a little bit faster. You may feel a rhythm beginning to occur in the work you do together. Or you may notice that one of the operations is lagging behind the rhythm. Perhaps there is a longer delay in receiving material from the preceding process or the next process seems to be coming sooner than before to pick up material from you.

By removing just one kanban you may be able to identify an area that could be improved. If so, then the improvement effort should immediately be focused on that area until the problem disappears. You can then remove another kanban and pay attention to the next problem area that shows itself. Progressively, very gradually, you chip away at hidden waste in this way, by reducing kanbans one at a time. After a while you will notice a vast improvement in the rhythm and performance of the whole factory. The effects will accumulate until you can almost tap your feet to the beat of takt time.

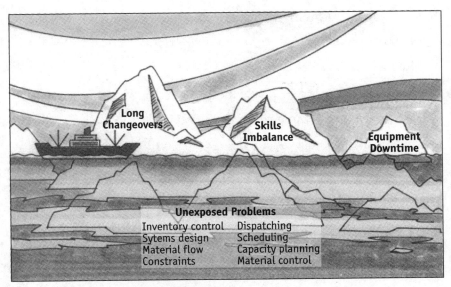

Figure 5-3. One Less Kanban Exposes Problems

Zero kanban is the ideal goal: The ultimate goal of this kanban reduction process is that when the downstream process withdraws parts, the inventory level at the product store of the upstream process would be zero and the next replenishment of the withdrawn material would occur immediately. This is an *ideal state* and difficult to achieve, and is not even practically possible for some types of products. But by reducing the number of kanbans one by one and addressing the problems that emerge as you do so, you can approach and come as close as possible to this ideal goal of *zero kanbans*.

Practicing changeovers: Assuming that the skills of all operators have been improved so that everyone knows their job and can do it to takt time, machine changeovers and setup times will emerge as the first level of problems when kanbans are reduced. Practicing changeovers is an important way to use idle time or designated improvement times. Once changeovers are reduced to a minimum (one minute if possible), and autonomous maintenance solutions help sustain machine uptime, deeper problems begin to emerge in the process. See Figure 5-3 for some things to pay attention to as the water level (number of kanbans/amount of stock carried) lowers.

To summarize, reducing kanbans:

- Exposes problems in the order of their priority

- Reduces safety stock

- Stimulates improvement activities

- Allows you to fine-tune your production cycles

- Let's you know how you are doing

- Helps you think about the right things

TAKE FIVE

Take five minutes to think about these questions and to write down your answers:

1. How much stock do you carry as a buffer in your cell or line?
2. What would happen if you reduced this amount?

Kanban As a Visual Control System

We have been talking about visual displays and visual controls throughout this book, and they have been discussed in many of the other shopfloor books as well. Needless to say, when a process becomes visible it is easier to follow and to improve. A visual factory is a clear indication that lean production is underway. 5S is the beginning of creating the visual factory. One-piece flow and cellular manufacturing take the visual concept further. In the kanban system this important focus on visual is taken to a new, more refined level. In this section we will discuss the importance of the visual aspect of the kanban system.

Key Point

The kanban system is a visual control system. It uses visual displays to communicate information about the status of the production process and it controls the process itself through the use of kanbans.

What Is the Essence of Control?

Consider the problem of control faced by a group of cowboys moving a herd of thousands of cattle hundreds of miles. What does control mean in this example? What if there were a regulation stating that every steer had to move in a straight line without wavering for those hundreds of miles? A cowboy would be needed to control the movement of each steer and even then it would not be possible to achieve. A conveyor belt could be built hundreds of miles long but what would be the point? The cattle move by themselves and it is their nature for the most part to stay in the herd. The cowboys are needed to point the herd in the right direction and to lasso and restrain them when they stray.

In other words, *controls are not needed when everything is normal.* They are only needed to point the way for the normal flow and to identify when abnormalities occur and quick action is needed. In the kanban system visual controls are used to keep a smooth flow of production going from customer order back to supplier purchasing, to limit the abnormalities that might occur without the visual controls, and to signal when abnormalities do occur.

Five Important Steps in Creating Visual Controls

The following are some basic applications of visual control that you will need for your kanban system:

1. Determine the locations where parts and products are to be stored between processes and mark them clearly. Indicate the storage location on the kanban. Figure 5-4 shows how signboards and lights alert material handlers (water beetles) to what material is needed and where it is located.

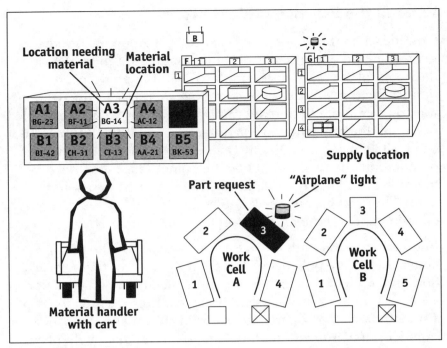

Figure 5-4. Storage Locations

2. Erect andons (display lamps and alarms) to signal when the line is stopped for defects or difficulties on the line, and as signals for replenishment of parts.

Figure 5-5. Andons

3. Place a kanban above the cell or production line to indicate what work is in process, the status of preparations, and so on.

Target	600
Actual	290
Rate	48.3
Defects	15

Figure 5-6. A Scheduling Board

4. Display kanbans so that cycle time, stock on hand, and work procedures can be known easily with a glance.

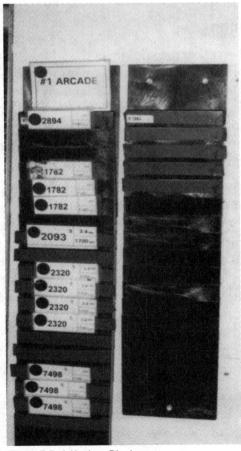

Figure 5-7. A Kanban Display

5. Include order-point indicators, or triangle kanbans (Figure 5-8), at the storage locations so that operators will know at a glance when to replenish. The next section discusses this step in detail.

TAKE FIVE

Take five minutes to think about these questions and to write down your answers:

1. How does the kanban system work as a visual control?

2. What types of visual controls do you have in your plant now?

3. What does it mean to *control* the process? Does the meaning of control discussed in this section change the way you think about lean production? How? What does it say about your role or your supervisor's role in a kanban system?

Triangle Kanban

Key Term

In Chapter 2 the reordering-point method of scheduling was discussed. *The reordering point is that level of inventory that must be maintained to insure that stock is available to meet average daily demand.* In a push system this level tends to be high and related to estimated monthly demand. In a pull system the goal is to reach the minimum level possible (zero stock being the ideal) and to keep it directly related to customer orders.

Key Point

When you begin to implement the kanban system your order point will probably be high. As you improve your system and reduce the number of kanbans circulating, the order point will decrease to a minimum amount of safety stock. *The triangle kanban system is a way of signaling when the order point is reached.* A triangle kanban is a special form of signal or production kanban. Figure 5-8 shows how the triangle kanban works to signal the order point in a storage system. When withdrawals are made down to the tagged point indicated by the triangle kanban, then a production kanban is circulated upstream to replenish the withdrawn parts.

Additionally, triangle kanbans are used to signal to the whole line when inventory levels have reached the order point. They are hung above the line next to the signs that indicate the product

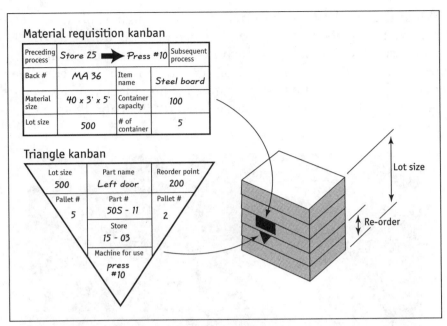

Material requisition kanban

Preceding process	Store 25 ➡ Press #10		Subsequent process
Back #	MA 36	Item name	Steel board
Material size	40 x 3' x 5'	Container capacity	100
Lot size	500	# of container	5

Triangle kanban

Lot size 500	Part name Left door	Reorder point 200
Pallet # 5	Part # 50S - 11	Pallet # 2
	Store 15 - 03	
	Machine for use press #10	

Lot size

Re-order

Figure 5-8. A Triangle Kanban As an Order-Point Indicator

being produced in that line or cell. The triangle kanban, when hung above the line, indicates that there are enough parts in inventory to produce the units currently being pulled by production kanbans. See Figure 5-9.

Figure 5-9. Triangle Kanbans as Above-the-Line Indicators

When the reordering point is reached the triangle kanban is removed from above the line and placed on a heijunka board, which sets the schedules for each line or machine, identifying how much and when the items on the kanbans will be made. See Figure 5-10. When the triangle kanban is no longer visible, everyone knows that the order point has been reached and that new parts will be made.

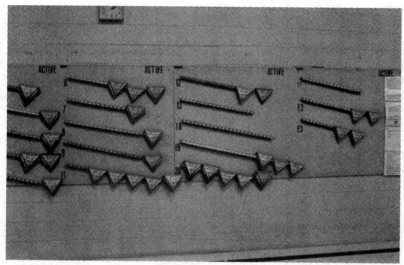

Figure 5-10. Triangle Kanbans on Heijunka Board Setting the Production Schedule

The roulette system and triangle kanban: The roulette system is helpful when parts are being used in amounts that are smaller than the standard pallet or container size. Eventually you will want to make improvements in the pallet size to match the quantity needed, but until then the roulette can be used with a triangle kanban to signal the reorder point. Figure 5-11 shows how the roulette system works. In that example, if a pallet or container holds sixty parts and no more than ten are needed in each shift, then there are enough parts for six shifts on one pallet. The roulette is divided into six parts, one for each shift. It is rotated clockwise, one block for each shift. When block 5 reaches the pointer at the bottom of the roulette, where block 2 now rests in Figure 5-11, a triangle kanban is placed in the production kanban post to signal the need to replenish the palette.

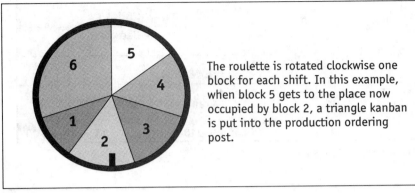

The roulette is rotated clockwise one block for each shift. In this example, when block 5 gets to the place now occupied by block 2, a triangle kanban is put into the production ordering post.

Figure 5-11. The Roulette System

TAKE FIVE

Take five minutes to think about these questions and to write down your answers:

1. Do you think triangle kanban would help in your plant? If so, why?
2. What is the reorder point for parts used in your operation?

Other Types of Kanbans

In addition to the basic types of kanbans described in Chapter 2 and explained throughout the book, there are a number of special kanbans that can be used to notify people when unusual circumstances occur.

Express kanban: An express kanban can be generated when there is an unexpected shortage of a part that the withdrawal and production kanbans did not address. This is to be used only in extraordinary situations and should be collected as soon as it has served its purpose. It is a withdrawal or production-ordering kanban with a bright red line across it to indicate that this is *express*; it may also be placed in a red kanban post for easy visibility.

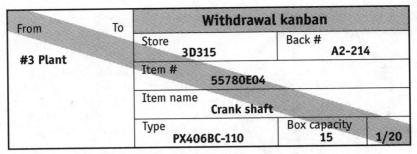

From	To	Withdrawal kanban		
		Store		Back #
#3 Plant		**3D315**		**A2-214**
		Item #		
		55780E04		
		Item name		
		Crank shaft		
		Type	Box capacity	
		PX406BC-110	**15**	**1/20**

Figure 5-12. An Express Kanban

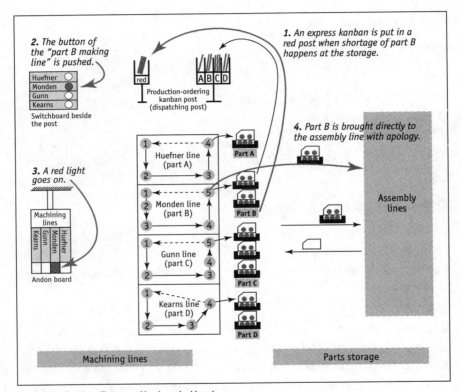

Figure 5-13. How Express Kanban Is Used

Emergency kanban: An emergency kanban signals that defective parts have been discovered and need to be replaced *immediately*. Like the express kanban, the emergency kanban should have its own distinctive look for priority handling. It, too, is collected as soon as the needed parts have been replaced.

Through kanban: If two or more processes work as one there is no need to exchange kanbans between these closely situated processes. In these cases a "through kanban" is used; for example, in machining lines where each piece is transferred to the next process through a chute, or in some other way is conveyed immediately to the next step. It is also used for process plants.

Common kanban: A withdrawal kanban can be used as a production kanban if two processes are close together and supervised by the same person.

Cart or truck as kanban: A cart is often the best way to transfer large parts to the next process or large subassemblies to final assembly. In these cases, the cart itself is the kanban. When the cart is empty it is taken back upstream to be filled again and returned to the production line. Although a kanban card is attached to each cart, it is the number of carts that indicates the status of production. If there is no empty cart then production stops. As soon as an empty cart appears, production begins again.

In some plants, raw materials are conveyed by truck and transferred for use as truckloads. An example would be scrap iron that is dumped into furnaces without being put into containers or counted in anyway. In this case the truck is considered as a kanban sheet or card.

Kanban as a label: Often a chain conveyor is used in final assembly to convey parts to the line. In final assembly a label is hung on the line to indicate which parts, how many, and when the parts are to be assembled. Labels like the one in Figure 5-14 are hung on a chain conveyor, which passes through the parts store where the required parts are hung on the conveyor and returned to final assembly. The label is also used to sequence the schedule of mixed models to be assembled.

AOI				
Assembly #		File #		Copy status
(Inter # ___th unit)		Destination		(Export car must use English plate)
Car style	BJ 43L – KJW			
Rear spring	Rear axle	Boster	Steering lock	Collapsible handle
	Semi	Single	Yes	/
Define gear ration	Free wheel facrication	Electric system	Exhaust	Transfer
411	/			Direct
Alternator	Air cleaner	Oil cooler	Heater & air conditioner	Front winch
480 w	/	/	Heater	/
Cold-climate oil	Altitude compensation	LLC	Fan	Rear hood
	/		Tempered	/
EDIC				Cold-climate destination
Yes				

Figure 5-14. Kanban as a Label for Final Assembly

Kanban innovations: In Figure 5-15 an ingenious type of kanban was created to eliminate the need to manually retrieve and issue withdrawal and production kanbans. When an assembly line worker opens a new box of parts, he or she removes the golf ball

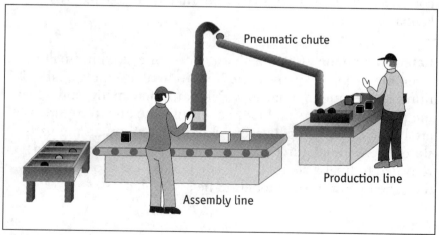

Figure 5-15. Golf Ball Kanban

kanban *that comes with the parts* and sends it through a pneumatic shoot to the production line. The shoot sends the balls through a gutter for 200 to 300 meters and plops them down in front of the production worker. They indicate different product models by their color, and quantities are written on the golf ball itself. They come in the order the parts were used so that production easily maintains the right order for replenishment.

TAKE FIVE

Take five minutes to think about these questions and to write down your answers:

1. Can you think of a new way to transfer information to your upstream process?

2. What types of situations occur in your plant that might require an express kanban?

Kanban and Suppliers

Eventually the kanban system will lead to the possibility of regulating orders from the factory to suppliers in a continuous chain starting with customer demand. When a plant reaches this level of kanban implementation, where the pull system is tied to its suppliers, it can celebrate the achievement of true lean production. Overproduction will have been eliminated and in-process inventory will be minimized; inventory and supplier relations will be managed in response to customer demand; and flexibility to customers' changing needs will be maximized. Figure 5-16 shows the flow of the kanbans back to the purchasing process. Once a factory has established the kanban system internally, the job of maintaining minimal inventories from suppliers becomes possible and stability in the relationship with suppliers can develop.

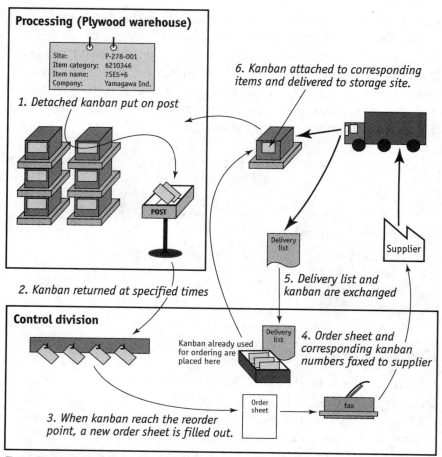

Processing (Plywood warehouse)

Site:	P-278-001
Item category:	6210346
Item name:	7SE5+6
Company:	Yamagawa Ind.

1. Detached kanban put on post

POST

6. Kanban attached to corresponding items and delivered to storage site.

Delivery list

Supplier

2. Kanban returned at specified times

5. Delivery list and kanban are exchanged

Control division

Kanban already used for ordering are placed here

Delivery list

4. Order sheet and corresponding kanban numbers faxed to supplier

Order sheet

fax

3. When kanban reach the reorder point, a new order sheet is filled out.

Figure 5-16. Purchasing Kanban for Suppliers

In Conclusion

SUMMARY

The kanban system reveals waste and problems very quickly. If setup times are too slow this will become evident when you try to work to takt time. Any process that is overproducing will have no room for the inventory they have produced because the pallet, bin, shelf, or supermarket designated for holding the material will be overflowing. Underproduction will also be evident by empty shelves, etc.

Reducing kanbans means reducing stock, which ultimately eliminates the safety stocks usually maintained as insurance against production fluctuations. Supervisors are given freedom to start with as many kanbans as they want and then, as the process continues, to reduce the number of kanbans one by one until the minimum possible is reached. In this way, the kanban system becomes a method of fine-tuning the production process. Reducing the number of kanbans being used causes problem areas to come out of hiding so that they can be improved. When you begin to implement the kanban system you start with the number of kanbans equal to current production flows.

Reducing kanbans to stimulate improvement activities is a separate process from changing the number of kanbans in response to customer demand. Reducing kanbans in response to customer demand is related to maintaining takt time as orders increase or decrease. These changes will not be dramatic and will be made as part of the scheduling process. It allows production to be completely flexible. Reducing the total number of kanbans in order to identify problem areas and reduce stock is different. Supervisors consciously reduce kanbans one by one to discover what the minimum number of kanbans can be in each production area.

This is a trial and error process and should be worked on as a team. Supervisors must pay attention to when the number of kanbans is too low. If the safety-stock level is too small, you will lose your motivation to improve. However, if the level is too high, work will become boring. As you use kanban reduction to improve your process, your work should become more enjoyable. This is a critical element to pay attention to as you improve.

The ultimate goal of this kanban reduction process is that when the downstream process withdraws parts, the inventory level at the product store of the upstream process would be zero and the next replenishment of the withdrawn material would occur immediately. This is an *ideal state* and difficult to achieve, and is not even practically possible for some types of products. But by reducing the number of kanbans one by one and addressing the problems that emerge as you do so, you can approach and come as close as possible to this ideal goal of *zero kanbans*.

Assuming that the skills of all operators have been improved so that everyone knows their job and can do it to takt time, machine changeovers and setup times will emerge as the first level of problems when kanbans are reduced. Practicing changeovers is an important way to use idle time or designated improvement times. Once changeovers are reduced to a minimum (one minute if possible), and autonomous maintenance solutions help sustain machine uptime, deeper problems begin to emerge in the process.

When a process becomes visible it is easier to follow and to improve. A visual factory is a clear indication that lean production is underway. 5S is the beginning of creating the visual factory. One-piece flow and cellular manufacturing take the visual concept further. In the kanban system this important focus on visual is taken to a new, more refined level.

The kanban system is a visual control system. It uses visual displays to communicate information about the status of the production process and it controls the process itself through the use of kanbans. *Controls are not needed when everything is normal.* They are only needed to point the way for the normal flow and to identify when abnormalities occur and quick action is needed. Visual controls are used in the kanban system to keep a smooth flow of production going from customer order back to supplier purchasing, to limit the abnormalities that might occur without them, and to signal when abnormalities do occur.

The reordering point is that level of inventory that must be maintained to insure that stock is available to meet average daily demand. In a push system this level tends to be high and related to estimated monthly demand. In a pull system the goal is to reach the minimum level possible (zero stock being the ideal) and to keep it directly related to customer orders.

In addition to the basic types of production and withdrawal kanbans, there are a number of other special kanbans that can be used to notify people when unusual circumstances occur: triangle kanban, express kanban, emergency kanban, through kanban, common kanban, a cart or truck as kanban, and labels for final assembly.

Ultimately, kanban needs to be extended to a factory's suppliers. This involves another advancement of the kanban system and lean production itself.

REFLECTIONS

Now that you have completed this chapter, take five minutes to think about these questions and to write down your answers:

- What did you learn from reading this chapter that stands out as particularly useful or interesting?

- Do you have any questions about the topics presented in this chapter? If so, what are they?

- What additional information do you need to fully understand the ideas presented in this chapter?

Chapter 6

Reflections and Conclusions

A Kanban Implementation Summary

The Six Rules of the Kanban System

Rule 1: Downstream processes withdraw items from upstream processes.

Rule 2: Upstream processes produce only what has been withdrawn.

Rule 3: Only 100 percent defect-free products are sent to the next process.

Rule 4: Level production must be established.

Rule 5: Kanbans always accompany the parts themselves.

Rule 6: The number of kanbans is decreased gradually over time.

A Summary of the Steps for Kanban Implementation

Phase One: Scheduling Kanban

1. Determine how many kanbans you need.
2. Calculate takt time—the necessary production rate to meet customer demand.
3. Determine the optimum number of operators needed for each process:
   a. Create maps of the process.
   b. Create cycle-time charts for each operator.
   c. Balance the lines until all operators and machines achieve full work.
4. Establish level production (heijunka).

Phase Two: Circulating Kanban

1. When the assembly line uses parts, they place the associated kanbans in the withdrawal kanban post.
2. The carrier removes a withdrawal kanban from the withdrawal kanban post and takes it to the preceding process to replace the items that have been used in assembly.
3. The carrier then removes a production kanban from the pallet or container and puts it in the production kanban post for that process. The withdrawal kanban is placed on the replenished pallet, which is then transported back to the assembly line area.

4. The production kanban removed from the withdrawn pallet serves as a production order to produce the withdrawn items.

5. Empty pallets are placed in the designated space.

6. As units are produced they are placed with the production kanbans in a storage area located within or near the production area so that carriers from subsequent processes can withdraw them at any time.

7. Withdrawal kanbans are taken to the preceding process to replace parts or subassemblies used to produce the replacement items in the same manner as in step 2.

Phase Three: Improving with Kanban

1. Fine-tune production by gradually reducing the number of kanbans being used and improving the problem areas that emerge.

2. Create visual controls with the kanban system:

   a. Determine locations where parts and products are to be stored between processes and mark them clearly. Indicate the storage location on the kanban.

   b. Erect andons (display lamps and alarms) to signal when the line is stopped for defects or difficulties on the line, and as signals for replenishment of parts.

   c. Place a kanban above the cell or production line to indicate what work is in process, the status of preparations, and so on.

   d. Display kanbans so that cycle time, stock on hand, and work procedures can be known easily with a glance.

   e. Include order-point indicators, or triangle kanbans, at the storage locations so that operators will know at a glance when to replenish.

Reflecting on What You've Learned

An important part of learning is reflecting on what you've learned. Without this step, learning can't take place effectively. That's why we've asked you to reflect at the end of each chapter. And now that you've reached the end of the book, we'd like to ask you to reflect on what you've learned from the book as a whole.

Take ten minutes to think about the following questions and to write down your answers.

- What did you learn from reading this book that stands out as particularly useful or interesting?

- What ideas, concepts, and techniques have you learned that will be *most* useful to you during implementation of the kanban system? How will they be useful?

- What ideas, concepts, and techniques have you learned that will be *least* useful during implementation of the kanban system? Why won't they be useful?

- Do you have any questions about the kanban system? If so, what are they?

Opportunities for Further Learning

Here are some ways to learn more about the kanban system:

- Find other books, videos, or trainings on this subject. Several are listed on the next pages.

- If your company is already implementing kanban, visit other departments or areas to see how they are applying the ideas and approaches you have learned about here.

- Find out how other companies have implemented kanban. You can do this by reading magazines and books about manufacturing cells, lean manufacturing, or just-in-time manufacturing, and by attending conferences and seminars presented by others.

Conclusions

The kanban system is more than a series of techniques. It is a fundamental approach to improving the manufacturing process. We hope this book has given you a taste of how and why this approach can be helpful and effective for you in your work.

Additional Resources Related to the Kanban System

Books and Videos

Just-in-Time, Lean Manufacturing, and One-Piece Flow

Shigeo Shingo, A *Study of the Toyota Production System: From an Industrial Engineering Viewpoint* (Productivity Press, 1989). This classic book was written by the renowned industrial engineer who helped develop the key elements of the Toyota system's success.

Japan Management Association (ed.), *Kanban and Just-in-Time at Toyota: Management Begins at the Workplace* (Productivity Press, 1986). This classic overview book describes the underlying concepts and main techniques of the original lean manufacturing system.

Jeffrey Liker, *Becoming Lean: Inside Stories of U.S. Manufacturers* (Productivity Press, 1997). This book shares powerful first-hand accounts of the complete process of implementing cellular manufacturing, just-in-time, and other aspects of lean production.

Hiroyuki Hirano, *JIT Implementation Manual: the Complete guide to Just-in-Time Manufacturing* (Productivity Press, 1990). This two-volume manual is a comprehensive, illustrated guide to every aspect of the lean manufacturing transformation.

Hiroyuki Hirano, *JIT Factory Revolution: A Pictorial Guide to Factory Design of the Future* (Productivity Press, 1988). This book of photographs and diagrams gives an excellent overview of the changes involved in implementing a lean, cellular manufacturing system.

Taiichi Ohno, *Toyota Production System: Beyond Large-Scale Production* (Productivity Press, 1988). This is the story of the first lean manufacturing system, told by the Toyota vice president who was responsible for implementing it.

Iwao Kobayashi, *20 Keys to Workplace Improvement* (Productivity Press, 1995). This book addresses 20 key areas in which a company must improve to maintain a world class manufacturing operation. A five-step improvement for each key is described and illustrated.

Ken'ichi Sekine, *One-Piece Flow: Cell Design for Transforming the Production Process* (Productivity Press, 1992). This comprehensive book describes how to redesign the factory layout for most effective deployment of equipment and people; it includes many examples and illustrations.

The 5S System and Visual Management

Tel-A-Train and the Productivity Press Development Team, *The 5S System: Workplace Organization and Standardization* (Tel-A-Train, 1997). Filmed at leading U.S. companies, this seven-tape training package (co-produced with Productivity Press) teaches shopfloor teams how to implement the 5S System.

Productivity Press Development Team, *5S for Operators: Five Pillars of the Visual Workplace* (Productivity Press, 1996). This Shopfloor Series book outlines five key principles for creating a clean, visually organized workplace that is easy and safe to work in. Contains numerous tools, illustrated examples, and how-to steps, as well as discussion questions and other learning features.

Michel Greif, *The Visual Factory: Building Participation Through Shared Information* (Productivity Press, 1991). This book shows how visual management techniques can provide "just-in-time" information to support teamwork and employee participation on the factory floor.

Poka-Yoke (Mistake-Proofing) and Zero Quality Control

Productivity Press Development Team, *Mistake-Proofing for Operators: The ZQC System* (Productivity Press, 1997). This Shopfloor Series book describes the basic theory behind mistake-proofing and introduces poka-yoke systems for preventing errors that lead to defects.

Shigeo Shingo, *Zero Quality Control: Source Inspection and the Poka-Yoke System* (Productivity Press, 1986). This classic book tells how Shingo developed his ZQC approach. It includes a detailed introduction to poka-yoke devices and many examples of their application in different situations.

NKS/Factory Magazine (eds.), *Poka-Yoke: Improving Product Quality by Preventing Defects* (Productivity Press, 1988). This illustrated book shares 240 poka-yoke examples implemented at different companies to catch errors and prevent defects.

Quick Changeover

Productivity Press Development Team, *Quick Changeover for Operators: The SMED System* (Productivity Press, 1996). This Shopfloor Series book describes the stages of changeover improvement with examples and illustrations.

Shigeo Shingo, *A Revolution in Manufacturing: The SMED System* (Productivity Press, 1985). This classic book tells the story of Shingo's SMED System, describes how to implement it, and provides many changeover improvement examples.

Total Productive Maintenance

Japan Institute of Plant Maintenance (ed.), *TPM for Every Operator* (Productivity Press, 1996). This Shopfloor Series book introduces basic concepts of TPM, with emphasis on the six big equipment-related losses, autonomous maintenance activities, and safety.

Japan Institute of Plant Maintenance (ed.), *Autonomous Maintenance for Operators* (Productivity Press, 1997). This Shopfloor Series book on key autonomous maintenance activities includes chapters on cleaning/inspection, lubrication, localized containment of contamination, and one-point lessons related to maintenance.

Newsletters

Lean Manufacturing Advisor—News and case studies on how companies are implementing lean manufacturing philosophy and specific techniques such as kanban. For subscription information, call 1-800-394-6868.

Training and Consulting

Productivity Consulting Group offers a full range of consulting and training services on lean manufacturing approaches and kanban. For additional information, call 1-800-394-6868.

Website

Visit our web pages at www.productivityinc.com to learn more about Productivity's products and services related to kanban.

About the Productivity Press Development Team

Since 1979, Productivity Press has been publishing and teaching the world's best methods for achieving manufacturing excellence. At the core of this effort is a team of dedicated product developers, including writers, instructional designers, editors, and producers, as well as content experts with years of experience in the field. Hands-on experience and networking keep the team in touch with changes in manufacturing as well as in knowledge sharing and delivery. The team also learns from customers and applies this knowledge to create effective vehicles that serve the learning needs of every level in the organization. to learn more about Productivity's products and services related to kanban.

About the Shopfloor Series

Put powerful and proven improvement tools in the hands of your entire workforce!

Progressive shopfloor improvement techniques are imperative for manufacturers who want to stay competitive and to achieve world class excellence. And it's the comprehensive education of all shopfloor workers that ensures full participation and success when implementing new programs. The Shopfloor Series books make practical information accessible to everyone by presenting major concepts and tools in simple, clear language. One main idea is presented every two to four pages so that the book can be picked up and put down easily. Each chapter begins with an overview and ends with a summary section. Helpful illustrations are used throughout.

Books currently in the Shopfloor Series include:

5S FOR OPERATORS
5 Pillars of the Visual Workplace
The Productivity Press Development Team
ISBN 1-56327-123-0 / 133 pages
Order 5SOP-BK / $25.00

QUICK CHANGEOVER FOR OPERATORS
The SMED System
The Productivity Press Development Team
ISBN 1-56327-125-7 / 93 pages
Order QCOOP-BK / $25.00

MISTAKE-PROOFING FOR OPERATORS
The Productivity Press Development Team
ISBN 1-56327-127-3 / 93 pages
Order ZQCOP-BK / $25.00

JUST-IN-TIME FOR OPERATORS
The Productivity Press Development Team
ISBN 1-56327-134-6 / 96 pages
Order JITOP-BK / $25.00

TPM FOR EVERY OPERATOR
The Japan Institute of Plant Maintenance
ISBN 1-56327-080-3 / 136 pages
Order TPMEO-BK / $25.00

TPM FOR SUPERVISORS
The Productivity Press Development Team
ISBN 1-56327-161-3 / 96 pages
Order TPMSUP-BK / $25.00

TPM TEAM GUIDE
Kunio Shirose
ISBN 1-56327-079-X / 175 pages
Order TGUIDE-BK / $25.00

AUTONOMOUS MAINTENANCE
The Japan Institute of Plant Maintenance
ISBN 1-56327-082-x / 138 pages
Order AUTOMOP-BK / $25.00

FOCUSED EQUIPMENT IMPROVEMENT FOR TPM TEAMS
The Japan Institute of Plant Maintenance
ISBN 1-56327-081-1 / 144 pages
Order FEIOP-BK / $25.00

OEE FOR OPERATORS
The Productivity Press Development Team
ISBN 1-56327-221-0 / 96 pages
Order OEEOP-BK / $25.00

CELLULAR MANUFACTURING
One-Piece Flow for Workteams
The Productivity Press Development Team
ISBN 1-56327-213-X / 96 pages
Order CELL-BK / $25.00

KANBAN FOR THE SHOPFLOOR
The Productivity Press Development Team
ISBN 1-56327-269-5 / 120 pages
Order KANOP-BK / $25.00

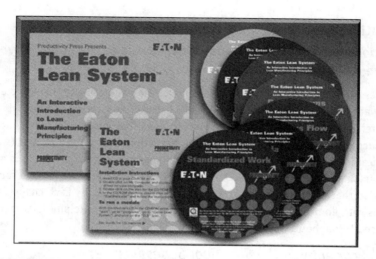

THE EATON LEAN SYSTEM
An Interactive Introduction to Lean Manufacturing Principles

If you're interested in a multi-media learning package, the best one available is *The Eaton Lean System*. Integrating the latest in interactivity with informative and powerful video presentations, this innovative software involves the user at every level. Nowhere else will you find the fundamental concepts of lean so accessible and interesting. Seven topic-focused CDs let you tackle lean subjects in the order you choose. Graphs, clocks and diagrams showing time wasted or dollars lost powerfully demonstrate the purpose of lean. Video clips show real people working either the lean way or the wasteful way. Easy to install and use, *The Eaton Lean System* offers the user exceptional flexibility. Either interact with the program on your own, or involve a whole group by using an LCD display.

This Software Package Includes:

7 CDs covering these important lean concepts!

- Muda
- Standardized Work
- Continuous Flow
- 5S (including 5S for administrative areas)

- Pull Systems
- Kaizen
- Heijunka

Includes in-plant video footage, interactive exercises and extensive simulations!

System Requirements — PC Compatible

Microsoft Windows® '98
8MB Free RAM

QuickTime for Windows 32 bit
16 bit display

The Eaton Lean System
The Productivity Development Team
ISBN 1-56327-261-X
Order EATON-BK / $695.00

Books from Productivity Press

Productivity Press publishes books that empower individuals and companies to achieve excellence in quality, productivity, and the creative
involvement of all employees. Through steadfast efforts to support the
vision and strategy of continuous improvement, Productivity Press delivers today's leading-edge tools and techniques gathered directly from
industry leaders around the world. Call toll-free (800) 394-6868 for our
free catalog.

KANBAN AND JUST-IN-TIME AT TOYOTA
Management Begins at the Workplace
Japan Management Association; Translated by David J. Lu
Toyota's world-renowned success proves that with kanban, the Just-In-Time production system (JIT) makes most other manufacturing practices obsolete. This
simple but powerful classic is based on seminars given by JIT creator Taiichi Ohno
to introduce Toyota's own supplier companies to JIT. It shows how to implement
the world's most efficient production system. A clear and complete introduction.
ISBN 0-915299-48-8 / 211 pages / $40.00 / Order KAN-BK

INTEGRATING KANBAN WITH MRPII
Automating a Pull System for Enhanced JIT Inventory Management
Raymond S. Louis
Manufacturing organizations continuously strive to match the supply of products
to market demand. Now for the first time, the automated kanban system is intro-
duced utilizing MRPII. This book describes an automated kanban system that integrates MRPII, kanban bar codes and a simple version of electronic data
interchange into a breakthrough system that substantially lowers inventory and
significantly eliminates non-value adding activities. This new system automatically
recalculates and triggers replenishment, integrates suppliers into the manufacturing loop, and uses bar codes to enhance speed and accuracy of the receipt
process. From this book, you will learn how to enhance the flexibility of your
manufacturing organization and dramatically improve your competitive position.
ISBN 1-56327-182-6 / 200 pages / $45.00 / Item # INTKAN-BK

JIT FACTORY REVOLUTION
A Pictorial Guide to Factory Design of the Future
Hiroyuki Hirano
The first encyclopedic picture-book of Just-In-Time, using photos and diagrams to
show exactly how JIT looks and functions in production and assembly plants.
Unprecedented behind-the-scenes look at multiprocess handling, cell technology,
quick changeovers, kanban, andon, and other visual control systems. See why a
picture is worth a thousand words.
ISBN 0-915299-44-5 / 218 pages / $50.00 / Order JITFAC-BK

TOYOTA PRODUCTION SYSTEM
Beyond Large-Scale Production
Taiichi Ohno

Here's the first information ever published in Japan on the Toyota production system (known as Just-In-Time manufacturing). Here Ohno, who created JIT for Toyota, reveals the origins, daring innovations, and ceaseless evolution of the Toyota system into a full management system. You'll learn how to manage JIT from the man who invented it, and to create a winning JIT environment in your own manufacturing operation.

ISBN 0-915299-14-3 / 163 pages / $45.00 / Order OTPS-BK

LEAN MANUFACTURING ADVISOR
Strategies and Tactics for Implementing TPM and Lean Production

What are others doing to implement lean or TPM? Anyone on a journey towards lean production asks themselves that question many, many times. Now, you can get the answers delivered to you every month in the *Lean Manufacturing Advisor*. Each issue brings you valuable news, advice, and the real-life, how to implement details from people on the same continuous improvement journey as you. We talk to and visit executives and managers who have experience in the trenches, so you can remove obstacles and speed implementation.

Lean Mfg Advisor / 12 monthly issues / $167.00 / Order LMA1YR-BK

BECOMING LEAN
Inside Stories of U.S. Manufacturers
Jeffrey Liker

Most other books on lean management focus on technical methods and offer a picture of what a lean system should look like. Some provide snapshots of before and after. This is the first book to provide technical descriptions of successful solutions and performance improvements. The first book to include powerful first-hand accounts of the complete process of change, its impact on the entire organization, and the rewards and benefits of becoming lean. At the heart of this book you will find the stories of American manufacturers who have successfully implemented lean methods. Authors offer personalized accounts of their organization's lean transformation, including struggles and successes, frustrations and surprises. Now you have a unique opportunity to go inside their implementation process to see what worked, what didn't, and why. Many of these executives and managers who led the charge to becoming lean in their organizations tell their stories here for the first time!

ISBN 1-56327-173-7/ 350 pages / $35.00 / Order LEAN-BK

ONE-PIECE FLOW
Cell Design for Transforming the Production Process
Kenichi Sekine

By reconfiguring your traditional assembly lines into production cells based on one-piece flow, you can drastically reduce your lead time, staffing requirements, and number of defects. Sekine examines the basic principles of process flow building, then offers detailed case studies of how various industries designed unique one-piece flow systems to meet their particular needs.
ISBN 0-915299-33-X / 308 pages / $75.00 / Order 1PIECE-BK

THE SAYINGS OF SHIGEO SHINGO
Key Strategies for Plant Improvement
Shigeo Shingo

Quality Digest calls Shigeo Shingo "an unquestioned genius—the Thomas Edison of Japan." Shingo offers new ways to discover the root causes of manufacturing problems. These discoveries can set in motion the chain of cause and effect, leading to greatly increased productivity." Hundreds of examples illustrate ways to identify, analyze and solve workplace problems.
ISBN 0-915299-15-1 / 207 pages / $45.00 / Order SAY-BK

MAKE NO MISTAKE!
An Outcome-Based Approach to Mistake-Proofing
C. Martin Hinckley

If you work for a company that emphasizes traditional quality control methods, it's unlikely that you've seen defects eliminated despite your substantial efforts. *Make No Mistake* clarifies the reasons why such traditional methods fail and shows how world-class quality can be achieved at a fraction of the cost through mistake-proofing – the practice of controlling virtually every source of potential errors.

Gathered here for the first time in a single source are the best methods for reducing complexity, variation, confusion, and the other root causes of defects. The centerpiece is an outcome-based classification system for mistake-proofing that focuses on preventing rather than detecting defects. When mistake-proofing practices are sorted according to specific outcomes, more similarities emerge in both the problems and the control methods than when organized by any other approach. Because mistake-proofing is a skill that improves through familiarity with previous solutions, this new classification system is the key to rapidly finding outstanding solutions to current problems on the shop floor.
ISBN 1-56327-227-X / 400 pages / $75.00 / Order MISTAKE-BK